JUMBO COLORING BOOK

COMMON NAME: ANTS

Ants are tiny bugs that live and work together in big groups. They build homes, find food, and help each other. Each ant has a job, and they are very good at working as a team.

Biographical Note
The Nature of Ants An Educational Coloring Book is a new work,
first published by Little Artist Studio in 2025.

International Standard Book Number
ISBN 979-8-9992504-7-6

www.littleartiststudio.org

Explore the incredible world of ants with over 90 detailed and captivating coloring pages. These industrious insects are essential to ecosystems, performing vital roles like soil aeration, seed spreading, and natural pest control. Part of Little Artist Studio's acclaimed educational coloring series, each full-page illustration is designed to spark curiosity, foster learning, and fuel creativity. Featuring single-sided pages, artists of all ages can use any medium and easily showcase their finished artwork. A perfect fusion of science and art for insect enthusiasts, educators, and curious minds of all ages.

SCIENTIFIC NAME FOR ANTS: FORMICIDAE

Ants belong to a big insect family called Formicidae. They are also part of a group of insects that includes bees and wasps (or Hymenoptera)!

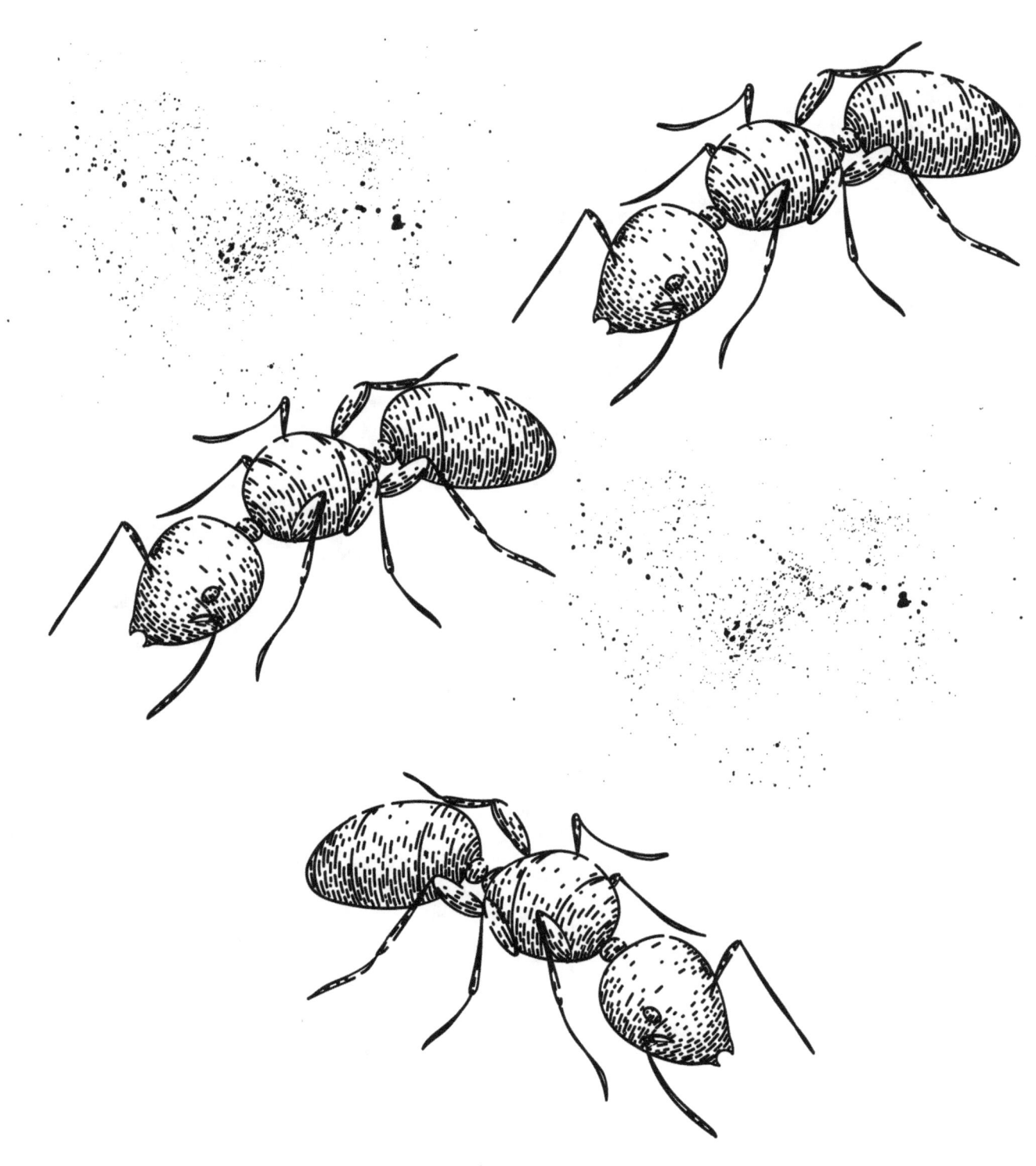

LEARNING ABOUT ANTS

Entomology is the study of insects, and ants are one type of insect. The ant life cycle has four main stages, just like wasps, bees and hornets!

STAGE 1: EGG

A tiny white dot laid by the queen.

STAGE 2: LARVA

A worm-like baby ant that gets fed by worker ants.

STAGE 3: PUPA

The larva curls up and starts to change, kind of like a cocoon.

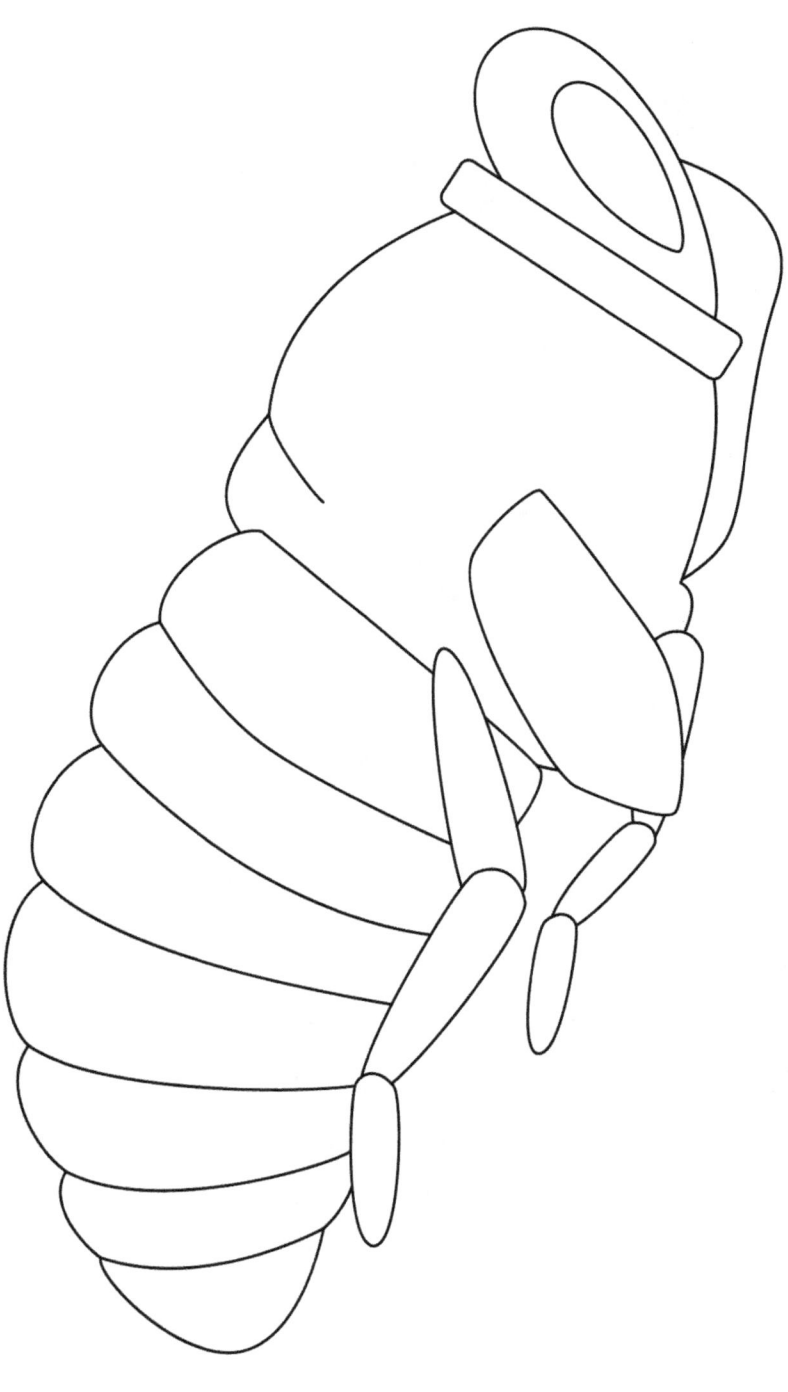

STAGE 4: ADULT

The ant is fully grown! It could be a worker, a soldier, or a new queen or male ant. Some ants live for a few weeks, and others can live for many years!

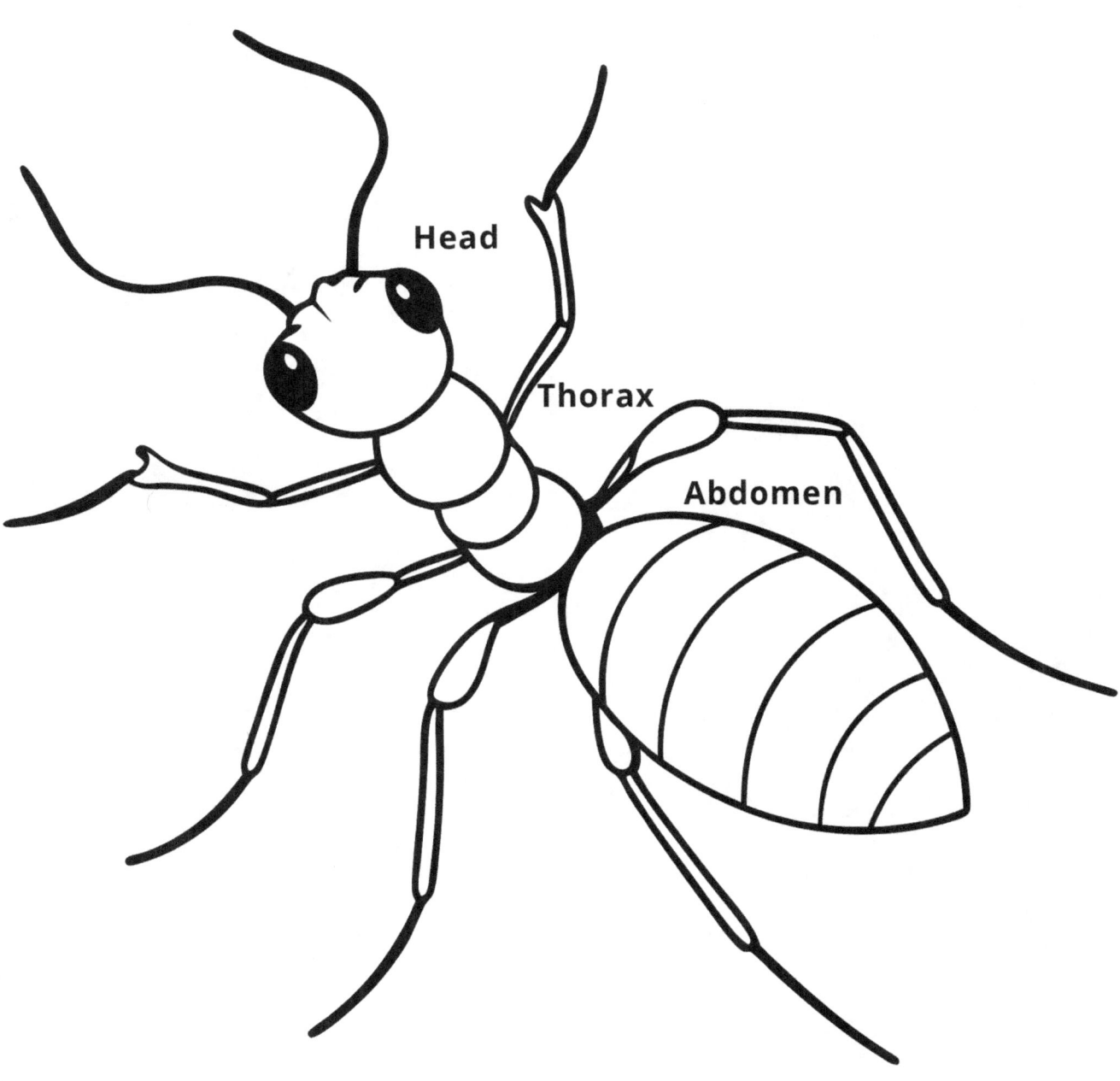

QUEEN ANT

Only queen ants lay eggs, and some ant queens can live for years!

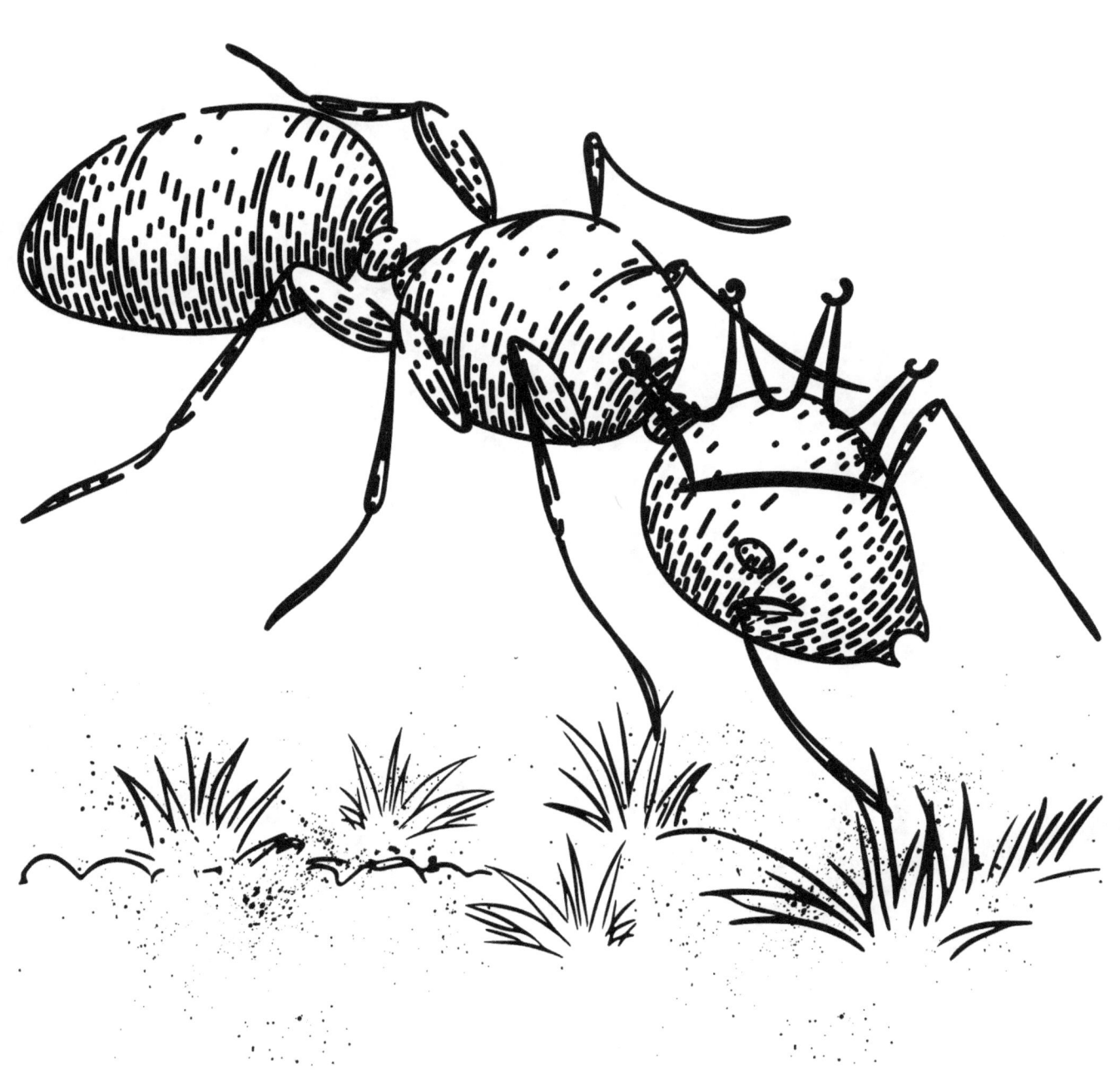

MALE ANTS

In some ant species, male ants (called drones) have one job - to mate with the queen. After that, they usually die.

WORKER ANTS

Worker ants are busy female ants that don't lay eggs. They gather food, take care of baby ants, build the nest, and keep the colony safe.

WORKER ANTS

Some worker ants are super strong-they can carry things 50 times heavier than themselves!

WINGED ANT

Ants with wings are special! They are either queens or boy ants called drones, and their job is to help start new ant families.

ANT VIBRATION

Ants don't have ears - they "hear" by feeling vibrations with their legs!

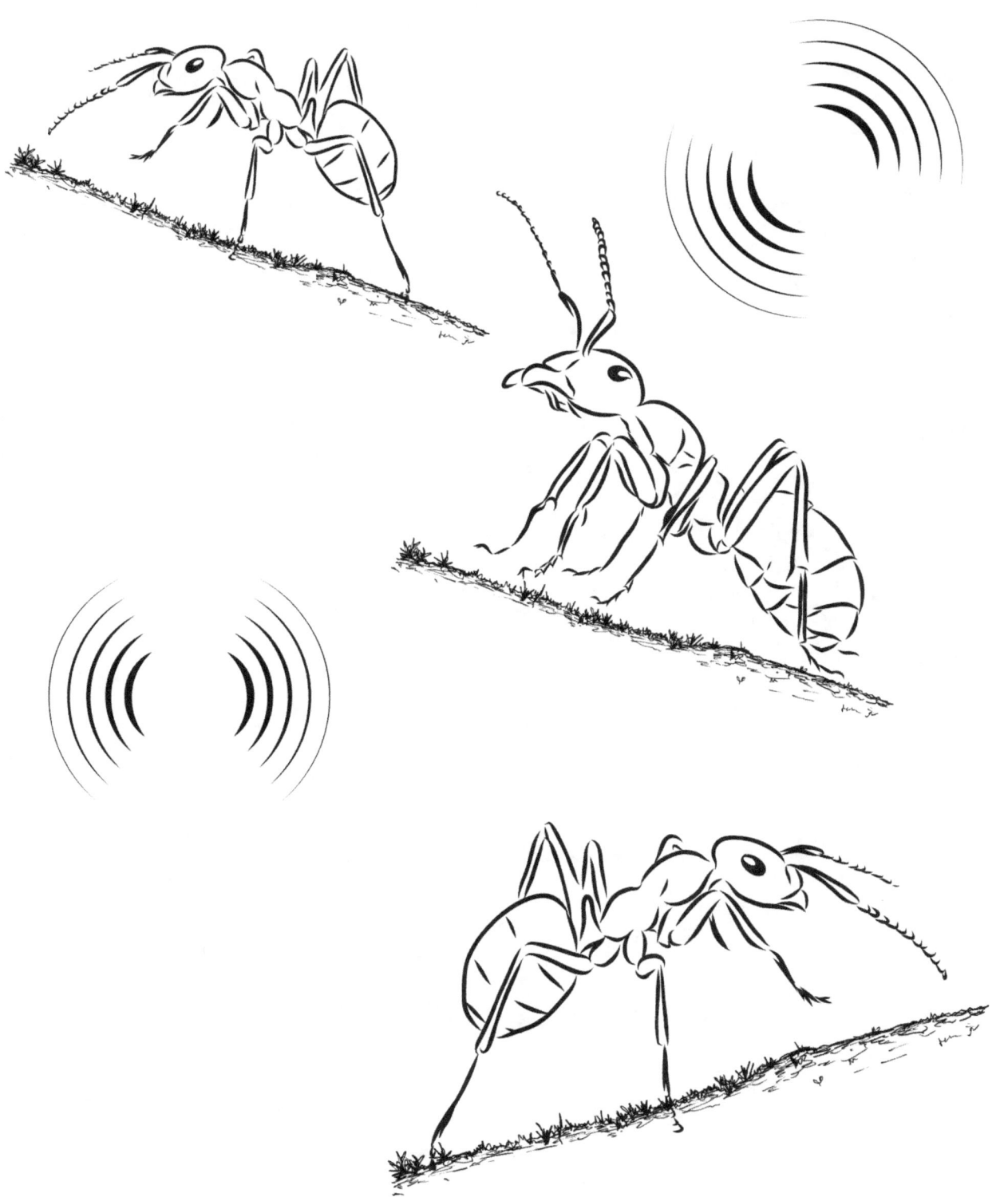

STRONG ANT

Ants are super strong for their size - they can carry 10 to 50 times their own body weight!

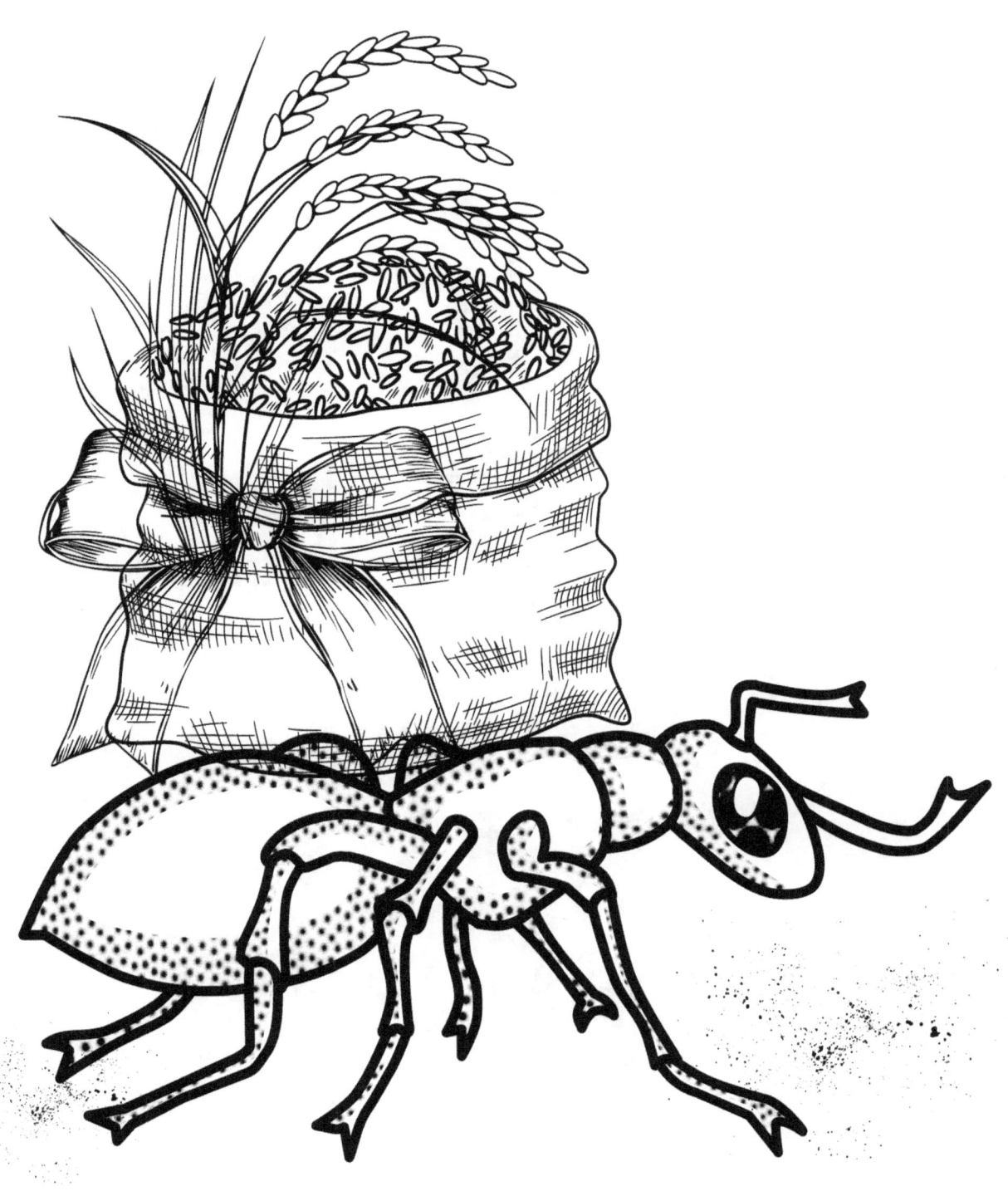

APPEARANCE

Ants come in all sizes-from teeny-tiny to about as long as your thumb! They can be black, brown, red, or even yellow-like tiny marching jellybeans!

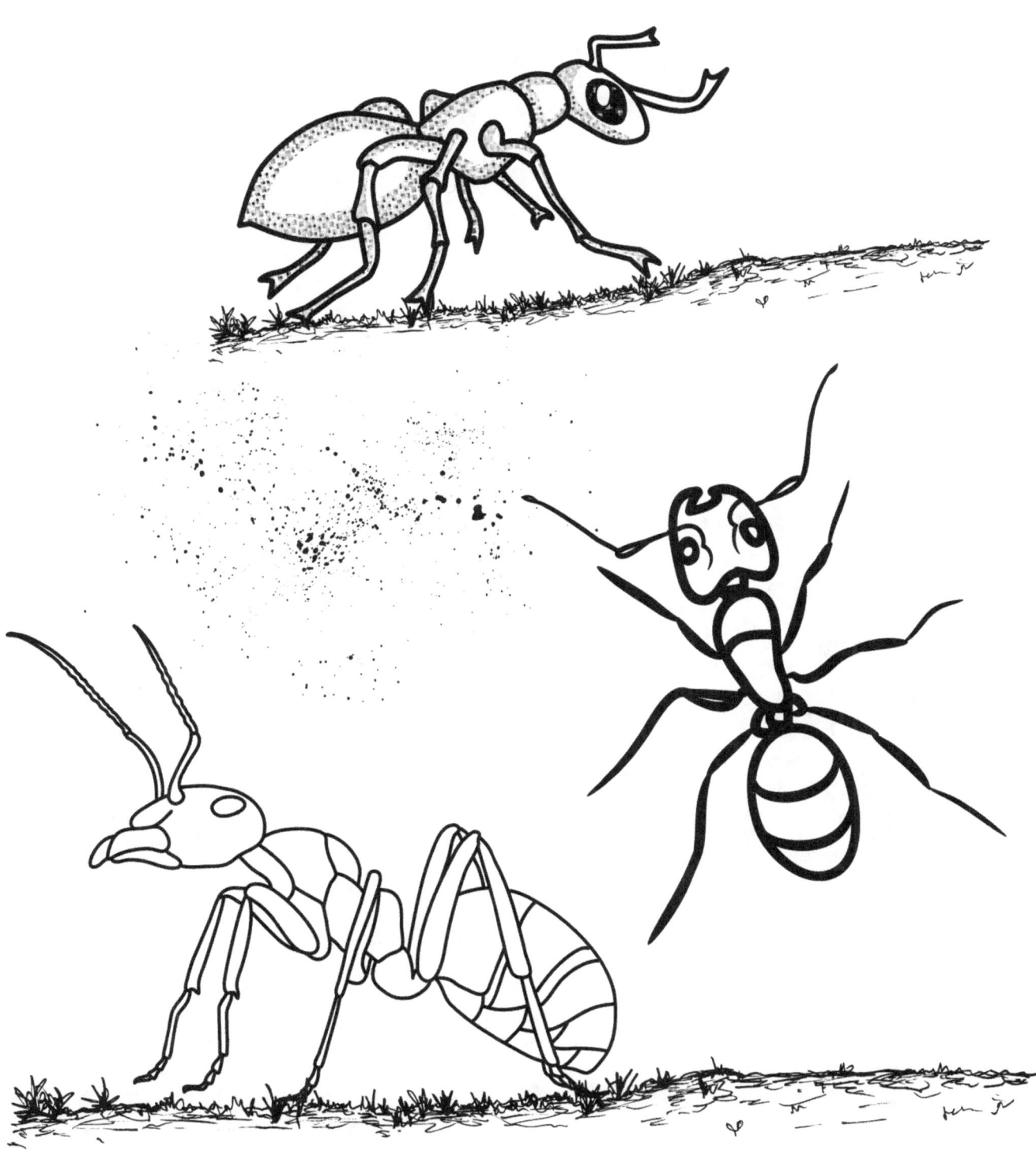

DIET: OMNIVORE

Ants usually eat things like nectar, dandelion seeds, fungus, or other insects.

ANTS NEST

Ants live together in busy homes called nests. Their nests can be underground, in big piles of dirt, or even up in trees!

ANT COLONY

A single ant colony can have hundreds of thousands of ants all living and working together like a big team!

ANT COLONIES

Some ant colonies have one queen, but others can have two or even thousands! The queens lay lots of eggs to help the colony grow and stay strong.

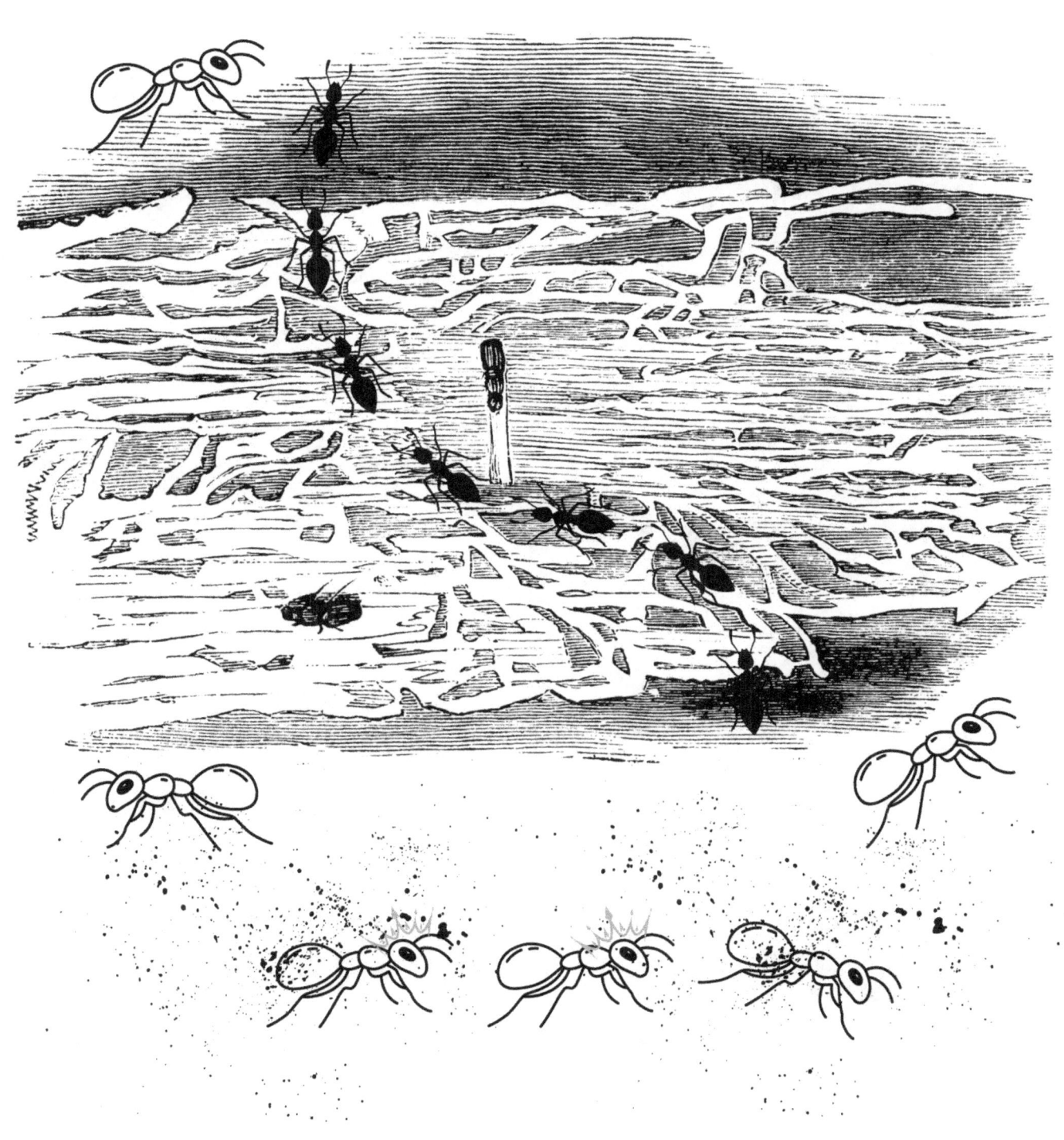

COLONY MEMORY

Ant colonies work so closely together that they can pass down helpful knowledge from one generation to the next - like a colony "memory"!

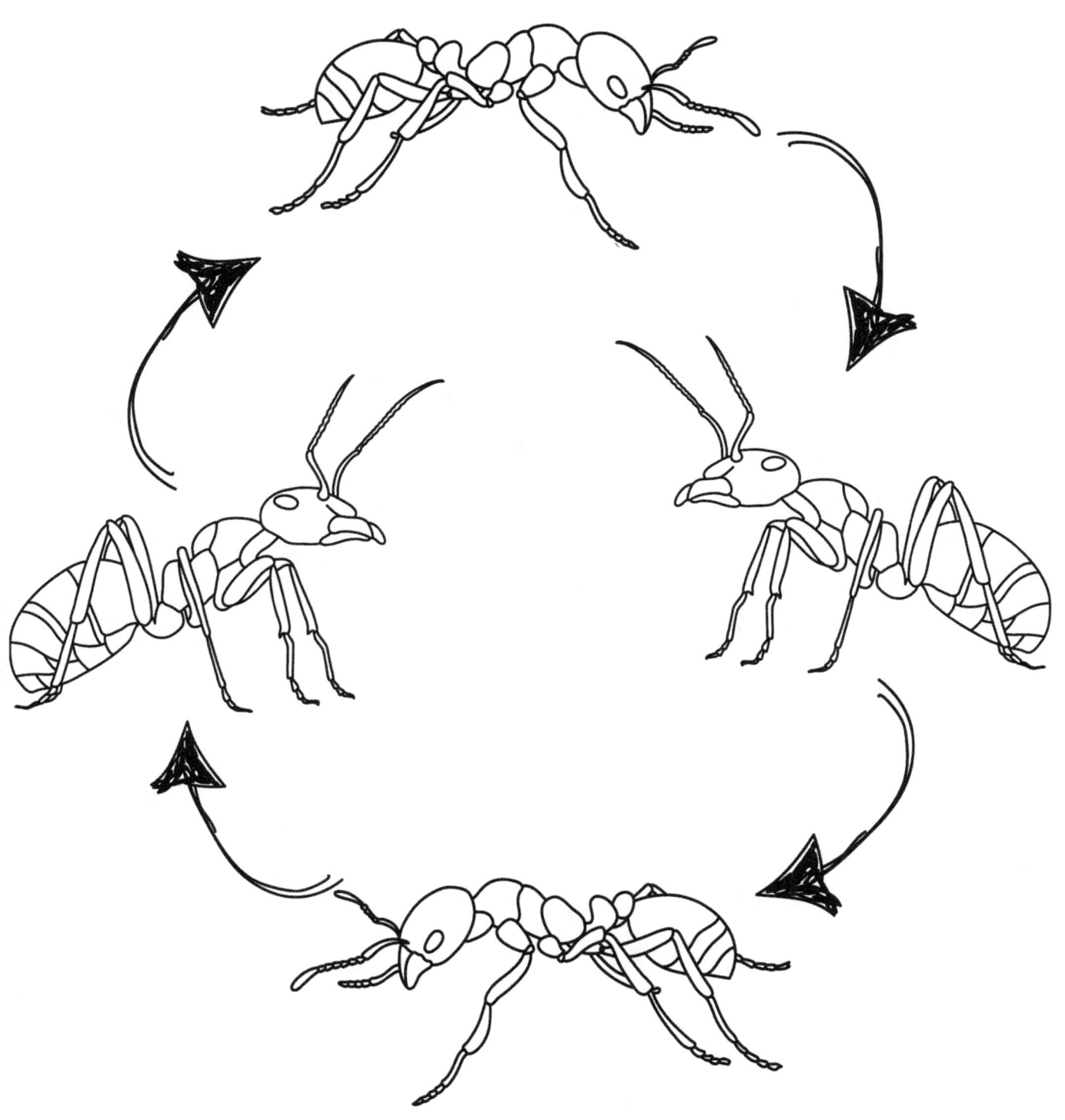

COLONY MEMORY

Colony memory helps ants know who is a friend and who might be a danger.

ANT OR TERMINTE?

Ants look a lot like termites, so people mix them up! But ants have bendy antennae and a skinny little waist-kind of like they're wearing a belt!

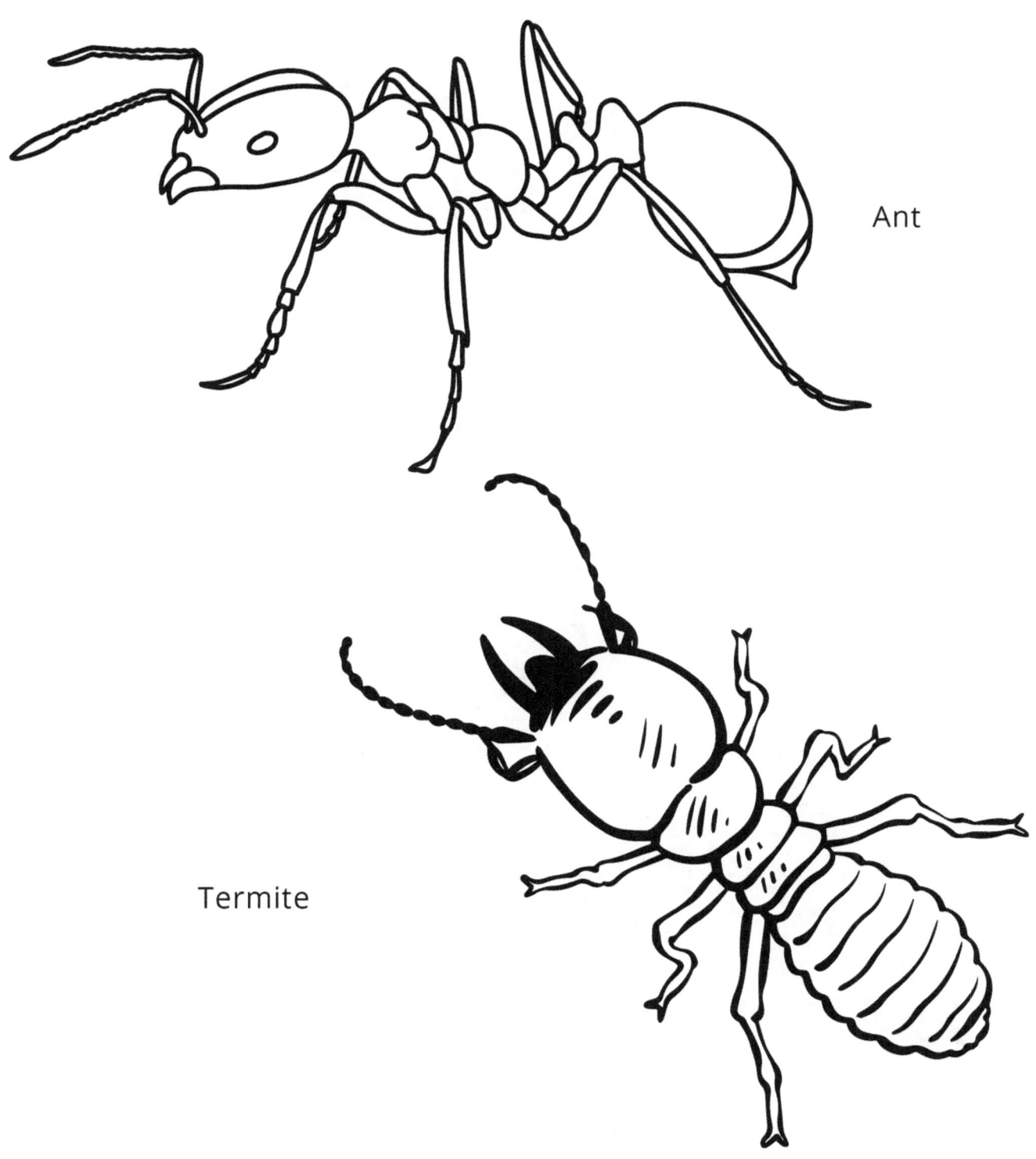

Ant

Termite

CARPENTER ANTS

Carpenter ants are a big ant family that lives in wood. Sometimes they make their homes in buildings, which can cause damage

ARMY ANTS

Some ants, like army ants, are always on the move and don't live in one place-they don't have a permanent home!

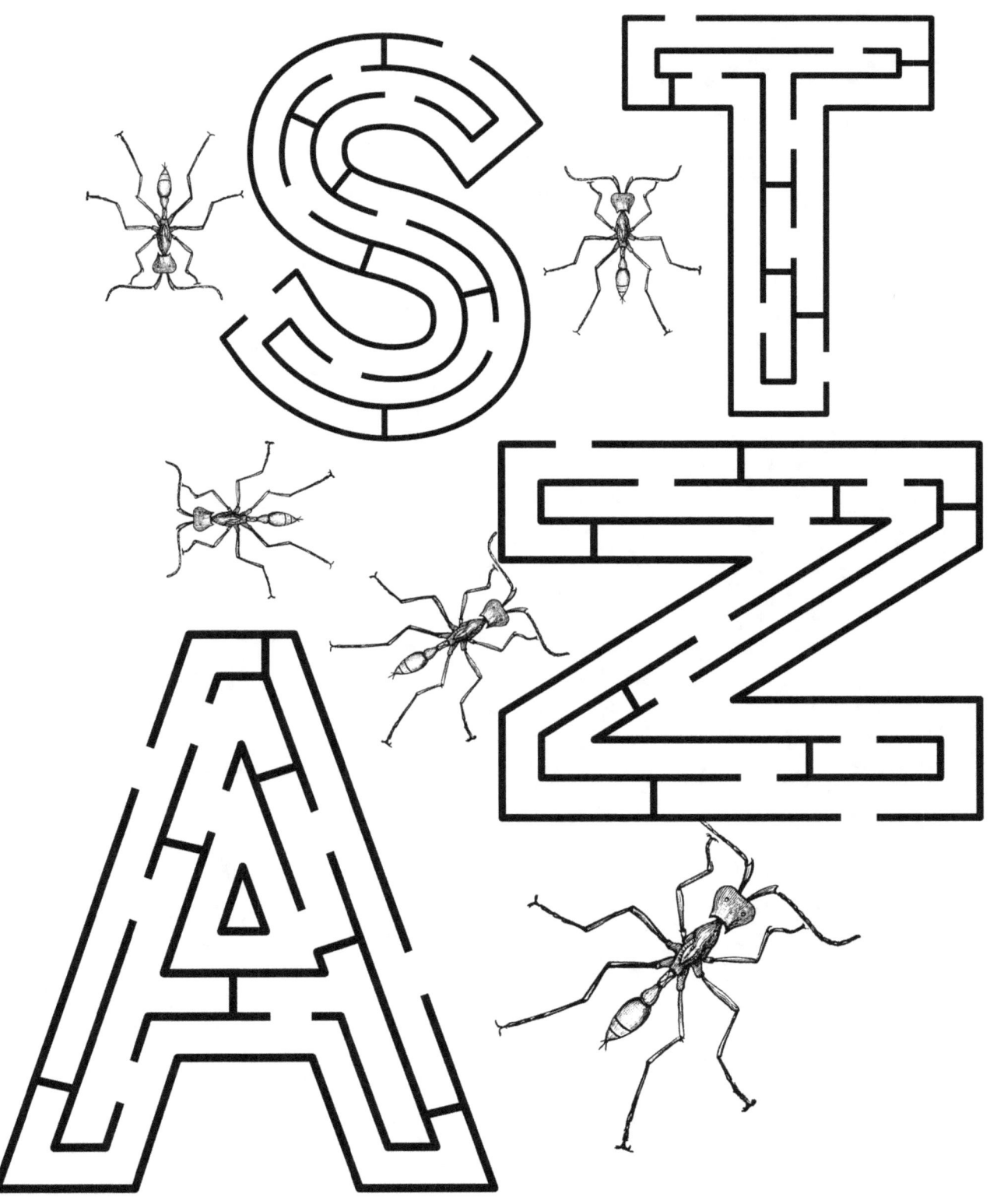

ARMY ANTS

Army ants get their name because they hunt like a team of soldiers. They work together in big groups to swarm an area and catch their food.

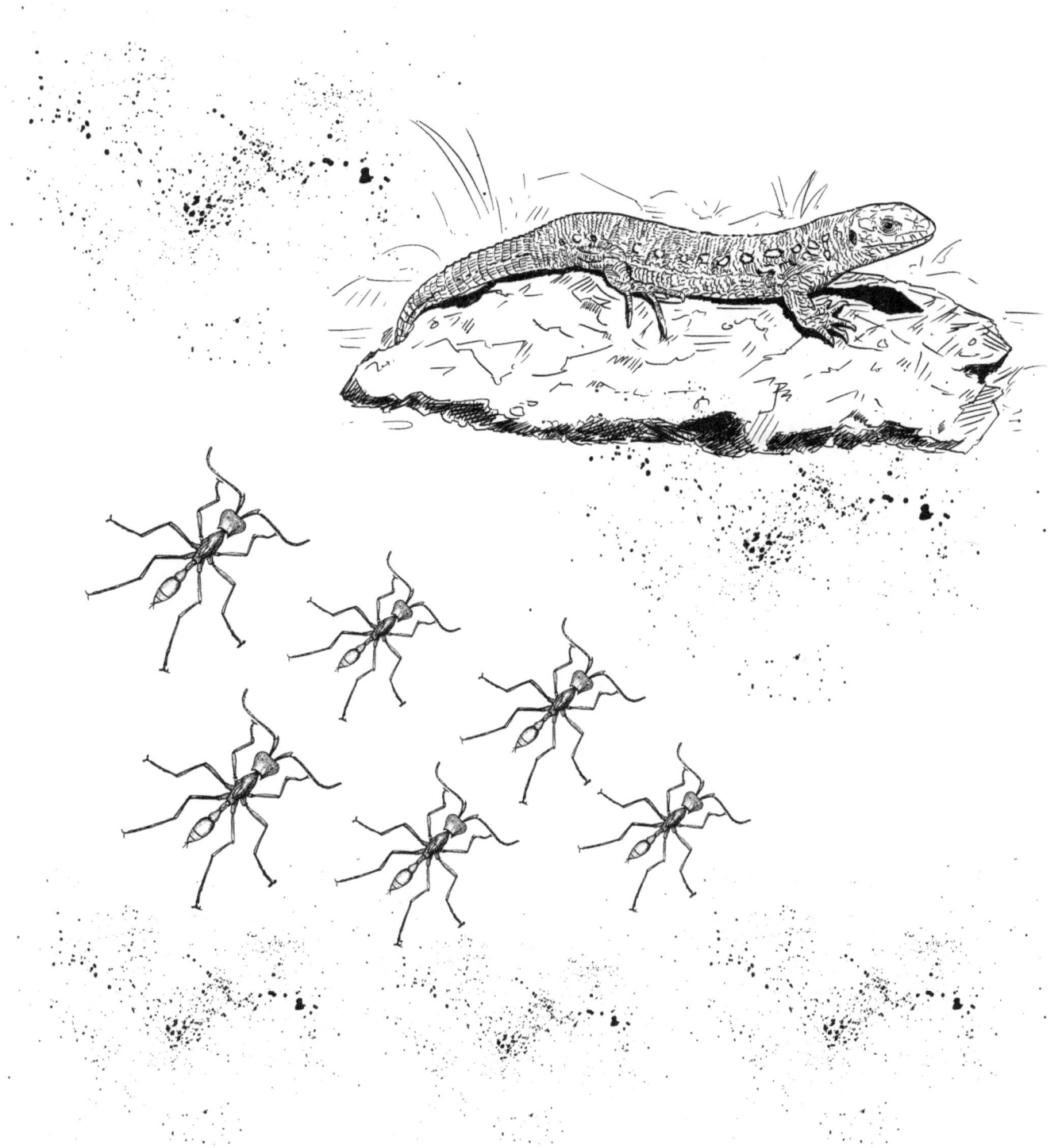

ANT COMMUNICATION

Ants talk to each other using special smells called pheromones. These scents help them find food or warn others about danger!

WHERE DO ANTS LIVE?

Ants live almost everywhere in the world-except in really cold places like Antarctica, Iceland, Greenland, and some islands.

LOCAL ECOSYSTEMS

Over 500 kinds of ants have shown up in places they don't belong, causing big problems for nature.

LOCAL ECOSYSTEMS

The red fire ant, native to South America, is known for its painful sting, bad temper, and harm to farms. When it shows up in new places, people got worried!

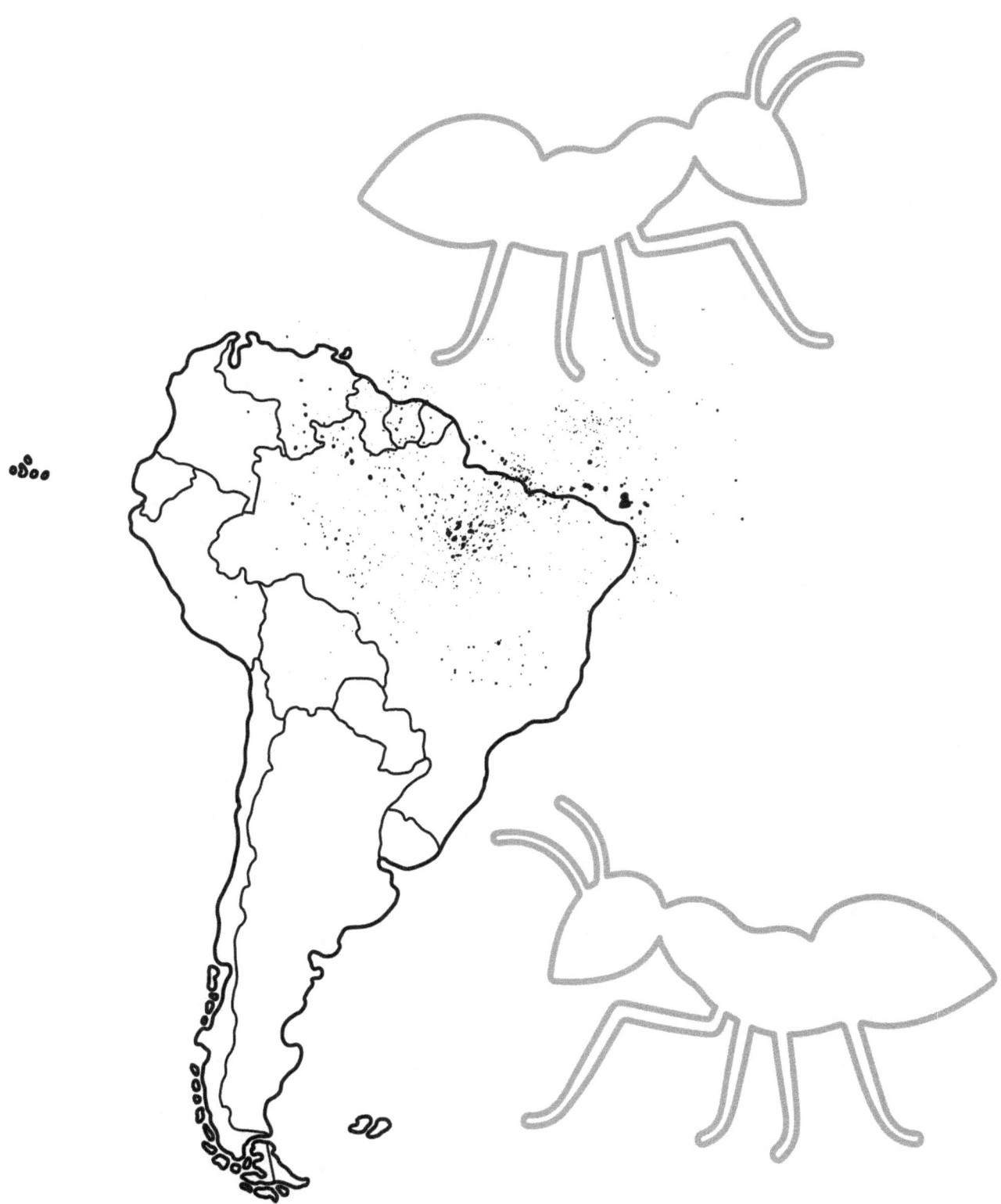

LOCAL ECOSYSTEMS

"Alien" ants like the red fire ant sometimes travel to new places by accident in cargo and packages. They don't belong there and can cause problems for nature.

LOCAL ECOSYSTEMS

An invasive fire ant called *Solenopsis geminata* lives in many warm places around the world. It can hurt nature and bother people with its painful sting.

BIOLOGICAL DEFENSES

Fire ants bite and sting using a venom called *solenopsin*, which can really hurt!

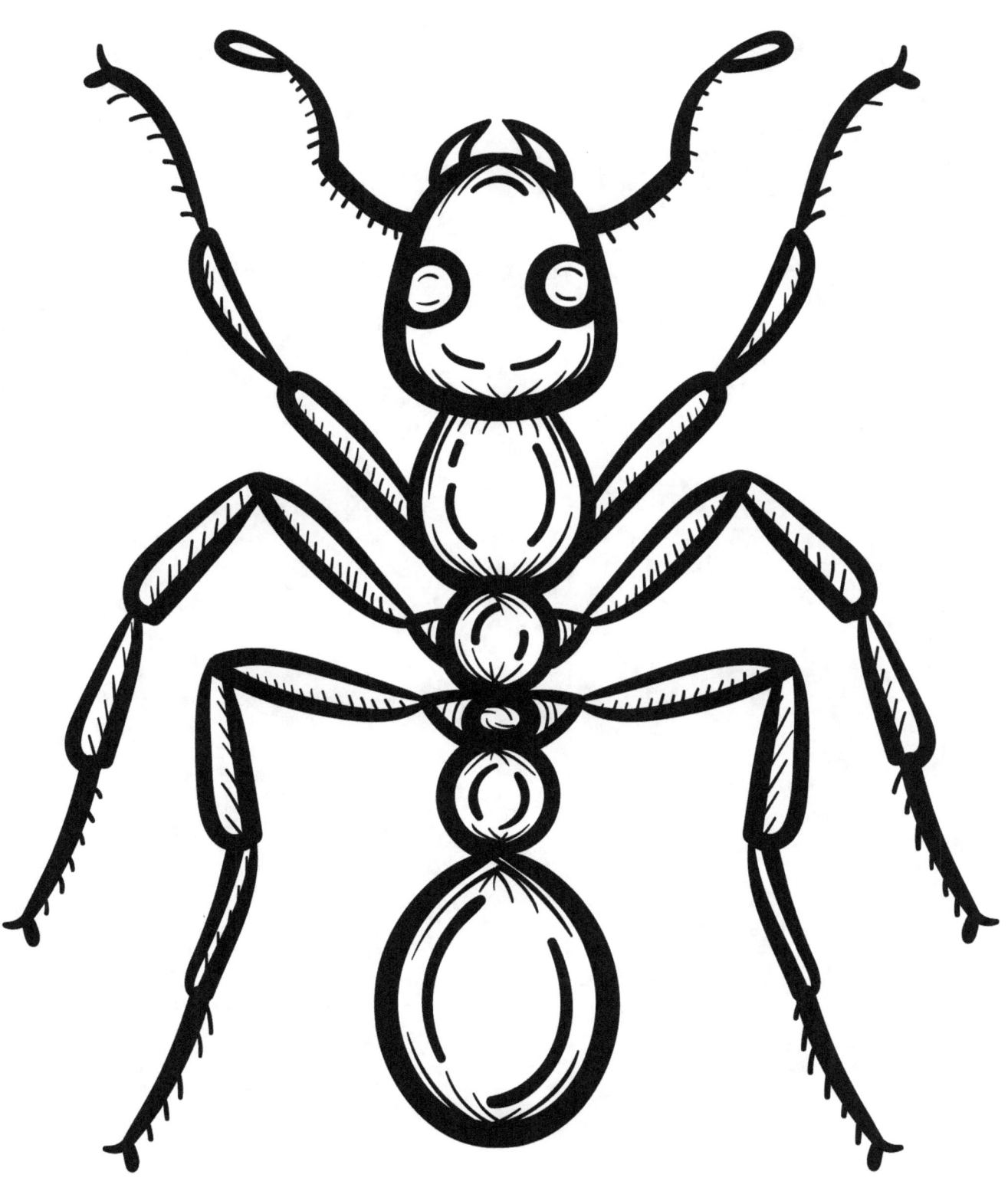

BIOLOGICAL DEFENSES

When there's a flood, fire ants hold onto each other to make a floating raft. Teamwork makes the ant-dream work!

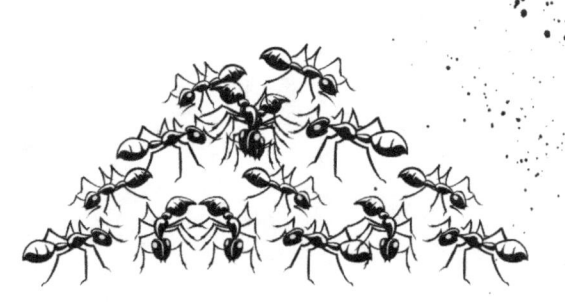

BIOLOGICAL DEFENSES

Some ants, like the newly found *Pheidole drogon*, have grown sharp spikes on their bodies to help protect themselves!

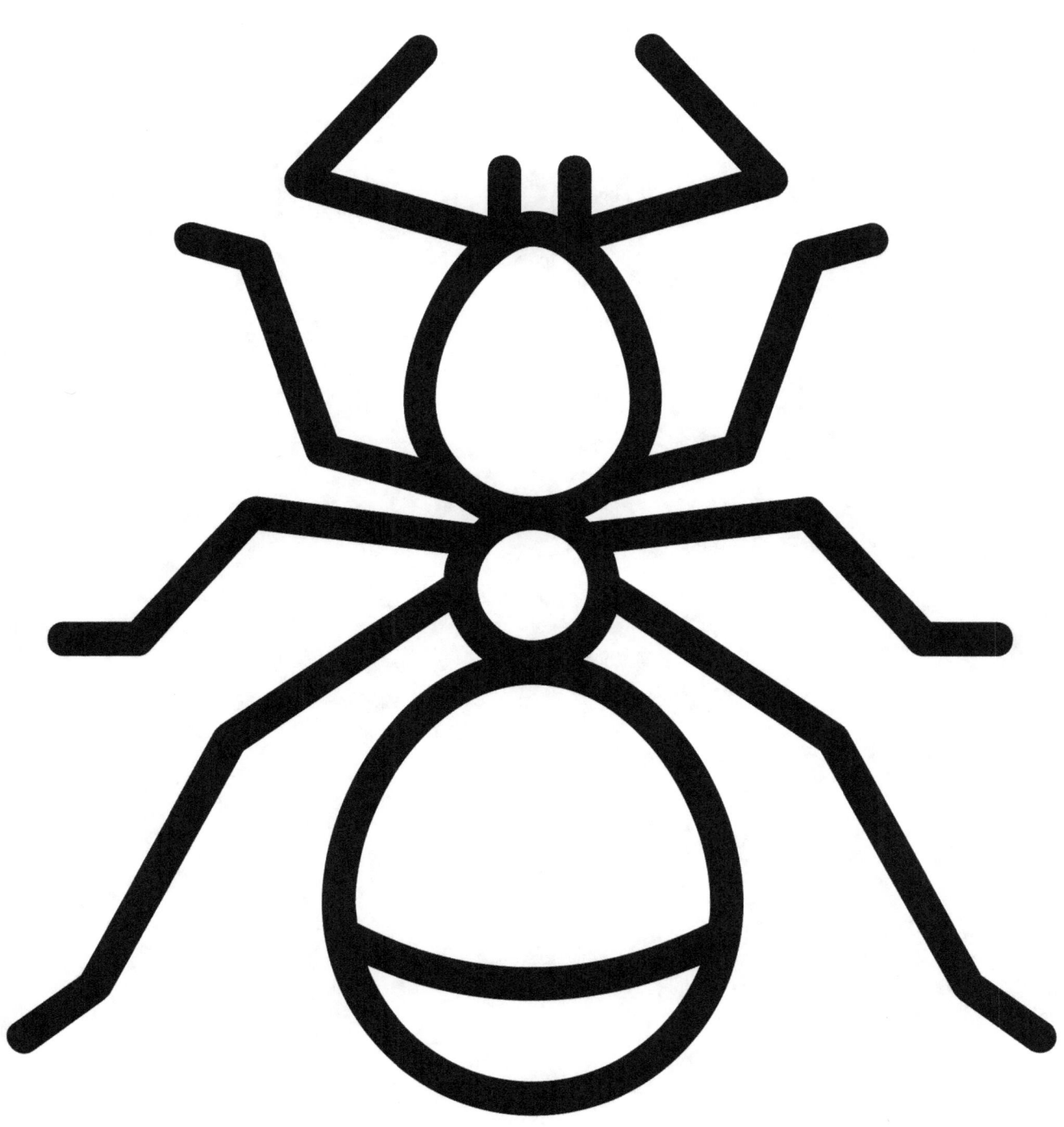

BIOLOGICAL DEFENSES

One Amazon ant, called *Allomerus decemarticulatus*, works together to build big traps out of plant pieces. When an insect steps on the trap, hundreds of ants grab it through tiny holes with their strong jaws!

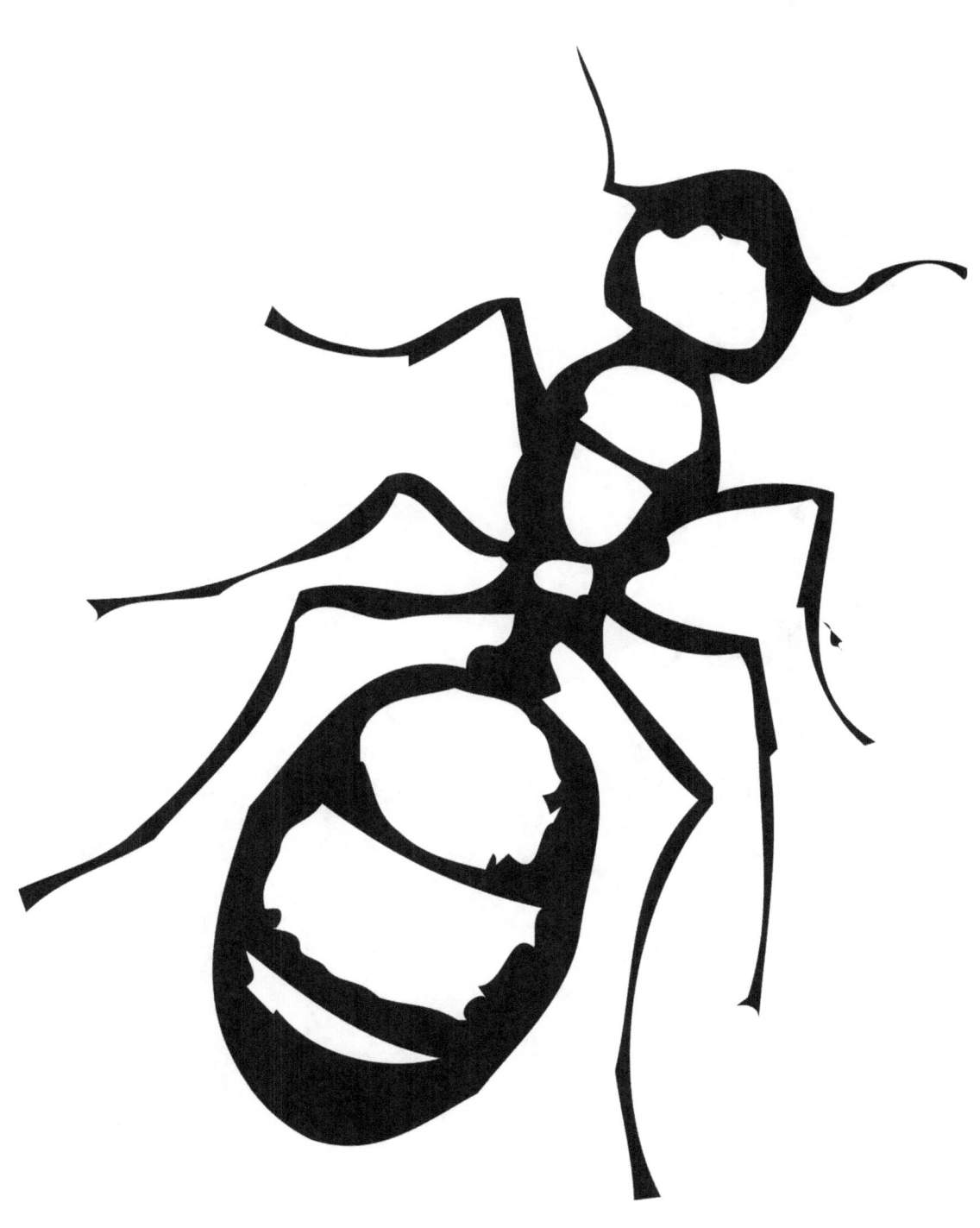

BIOLOGICAL DEFENSES

The yellow crazy ant can build giant homes called supercolonies with hundreds or even thousands of queens living together!

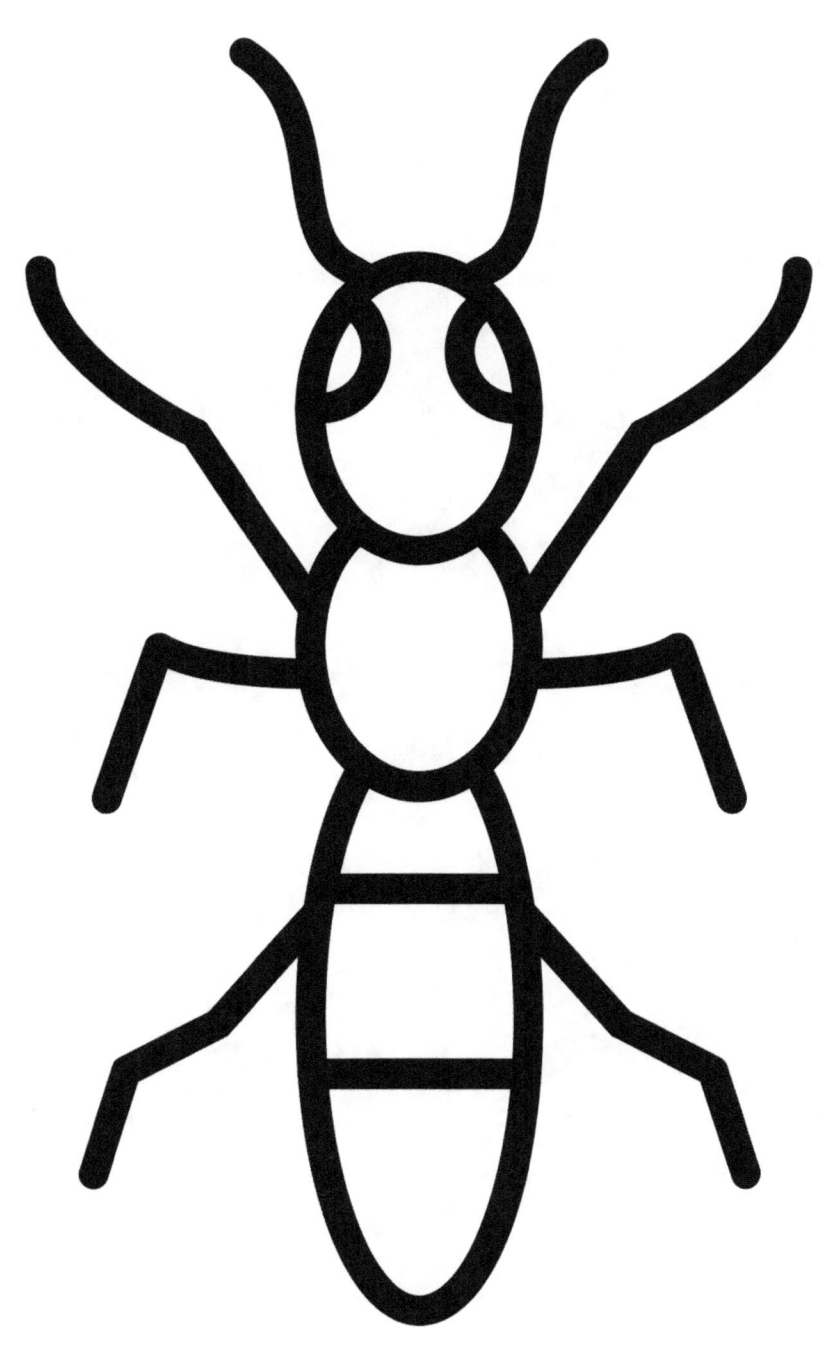

UNSUAL ANTS

There are thousands of different kinds of ants, and many have special jobs in nature.

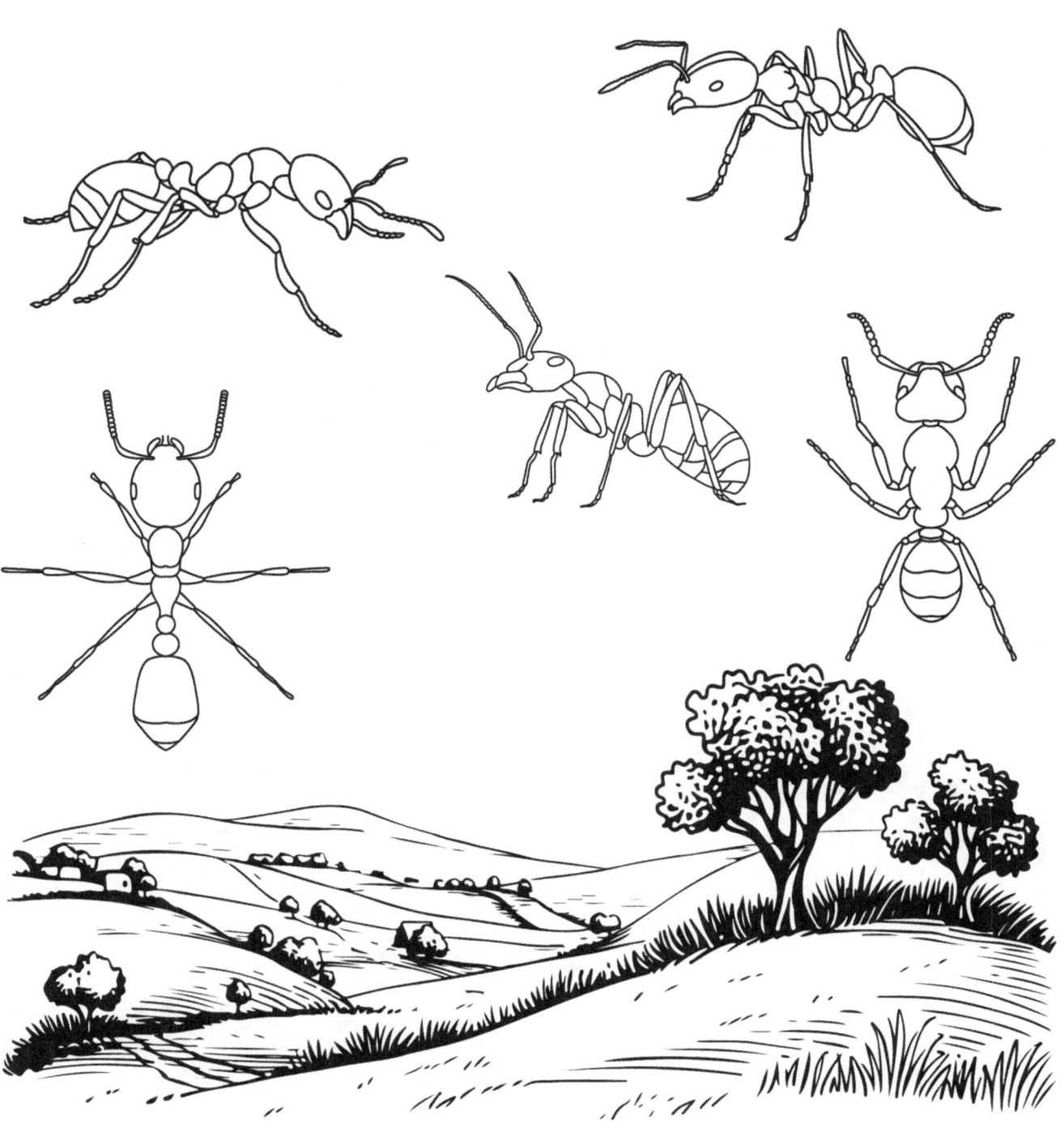

UNSUAL ANTS

On Australia's Christmas Island, yellow crazy ants were brought by accident, and now there are way too many, causing big problems for nature.

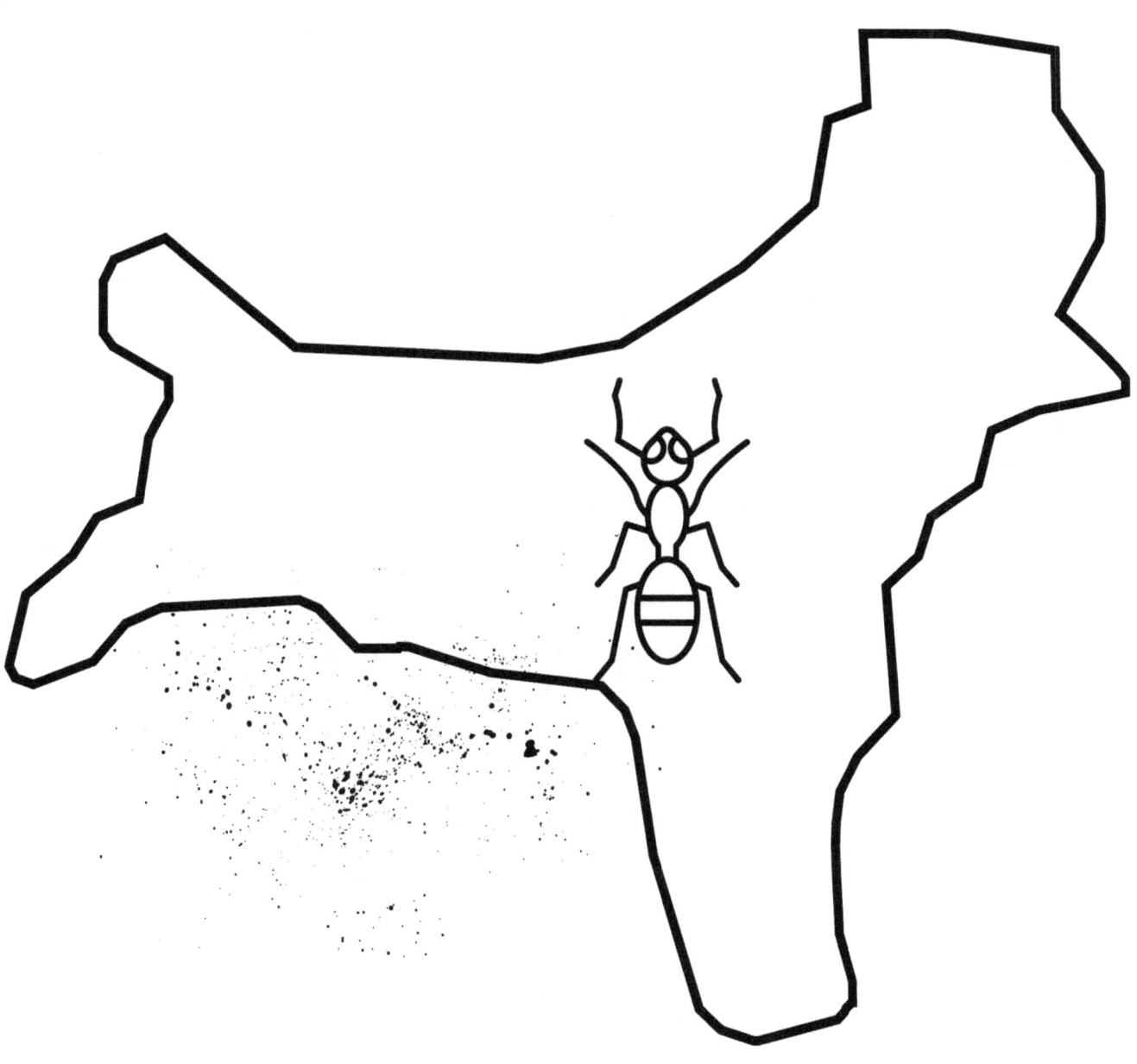

UNSUAL ANTS

Yellow crazy ants are a big danger to the Australian Christmas Island's red crab.

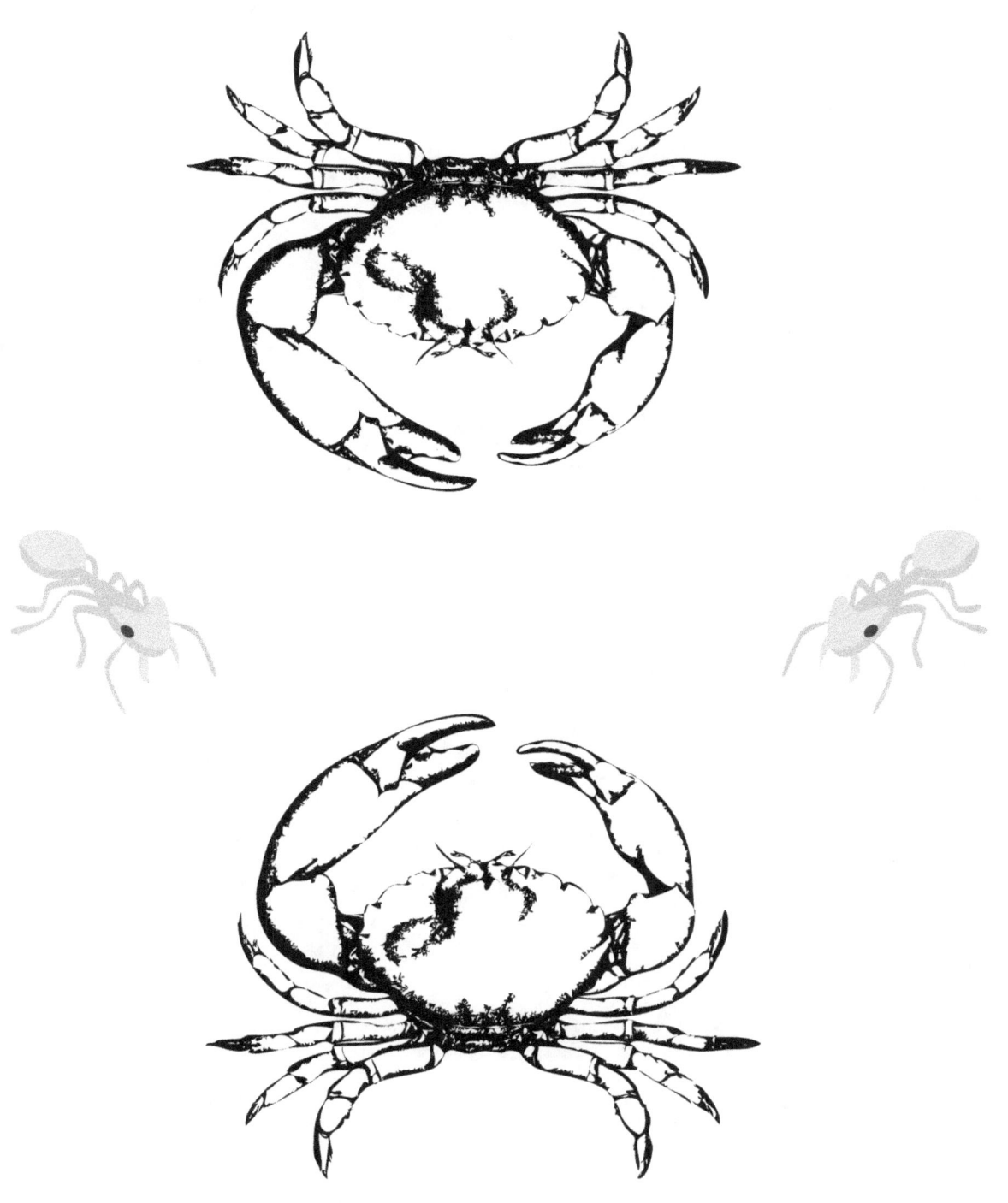

UNSUAL ANTS

Because of their jobs, some ants have cool body shapes and do interesting things!

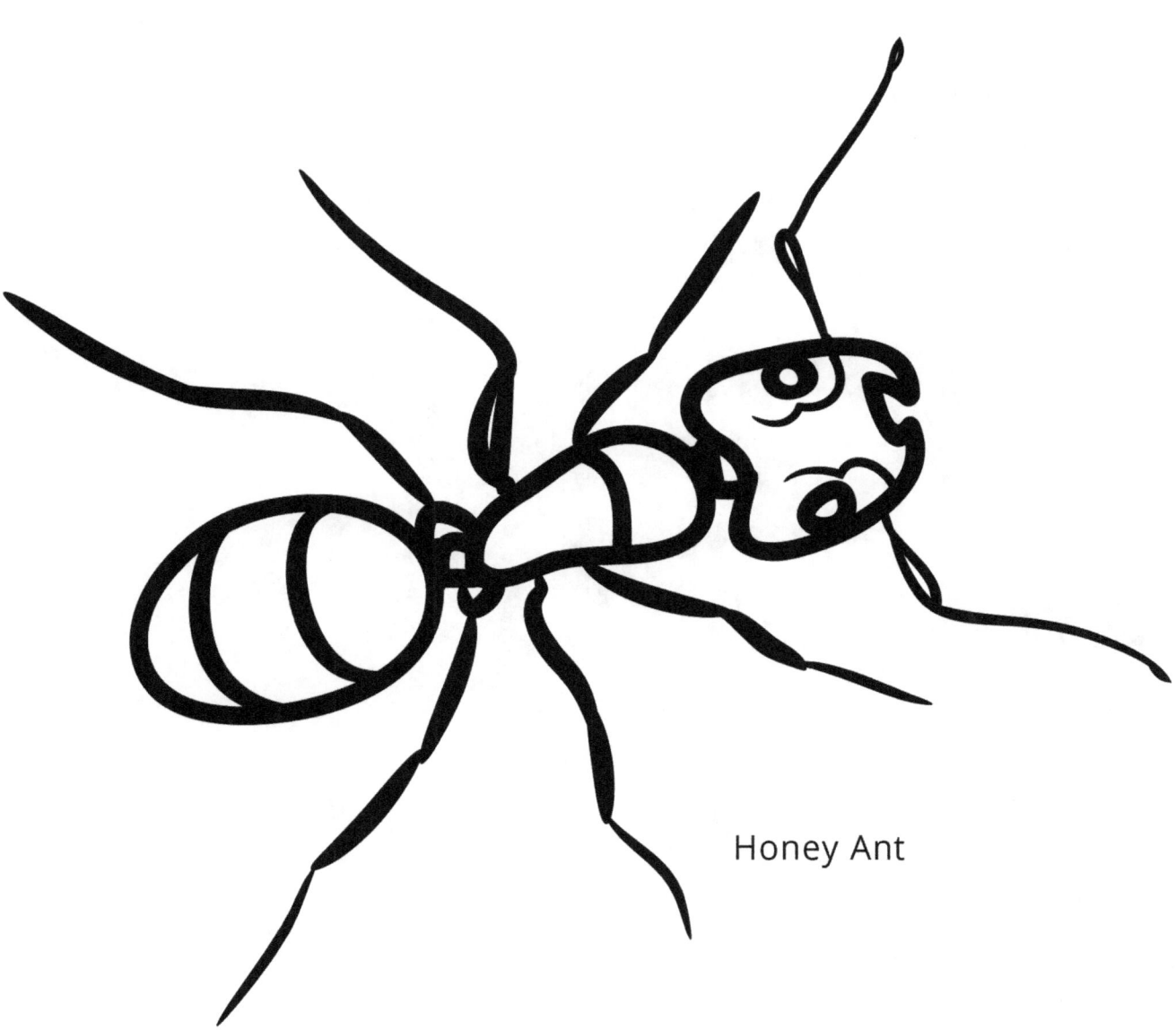

Honey Ant

DID YOU KNOW?

Some ants live inside the hollow parts of bamboo.The bamboo keeps them safe, and the ants help by chasing away bugs that might hurt the plant.

DID YOU KNOW?

The velvet ant is actually a wasp! The female velvet ants don't have wings, so they look like big, fuzzy ants.

DID YOU KNOW?

Scientists found a 113-million-year-old ant with sharp jaws that lived when dinosaurs roamed the Earth!

HONEYPOT ANT

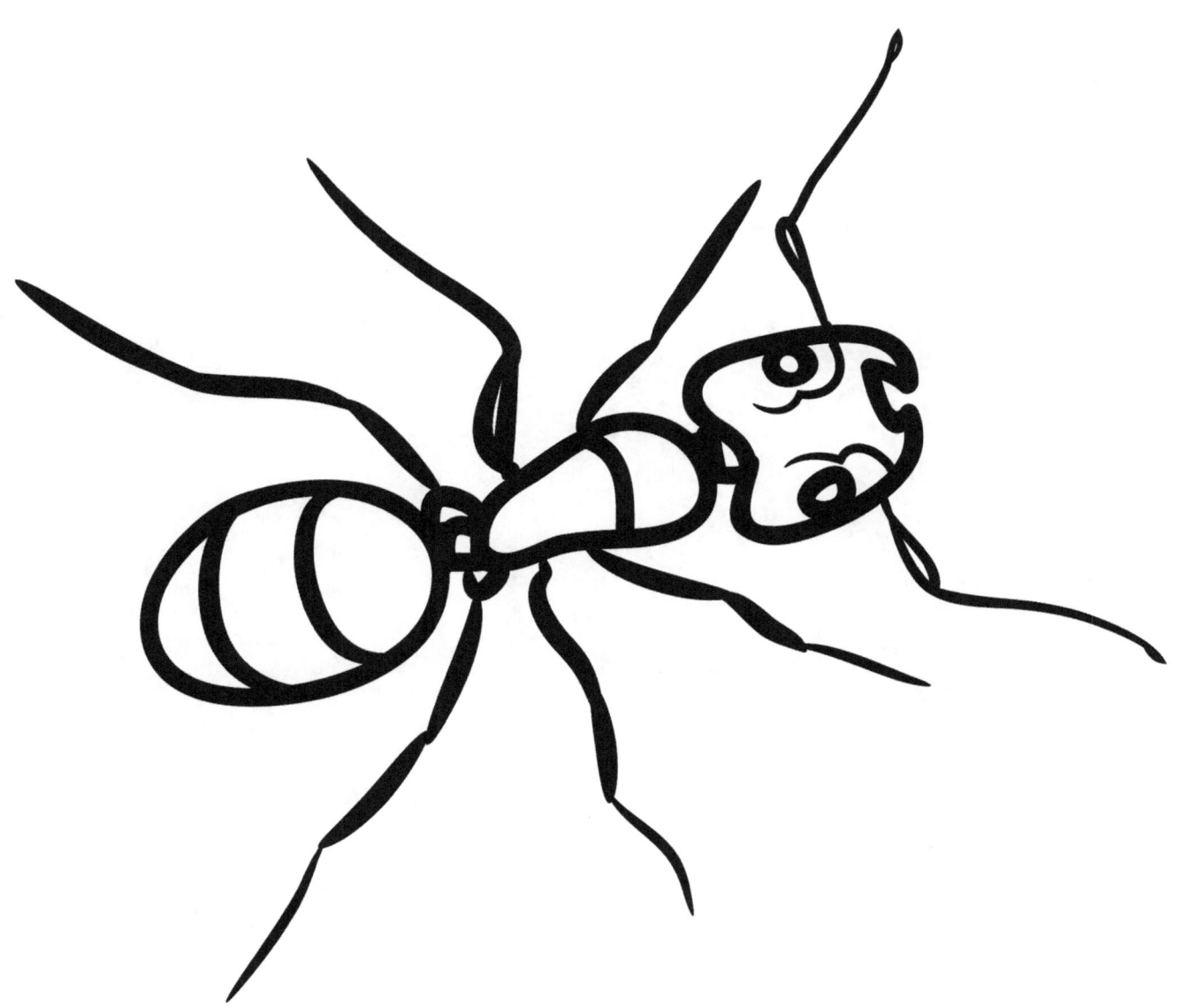

RED VELVET ANT

LEAF CUTTER ANT

MEGAPONERA ANALIS

Megaponera analis is a small black ant from Africa that loves to fight termite nests!

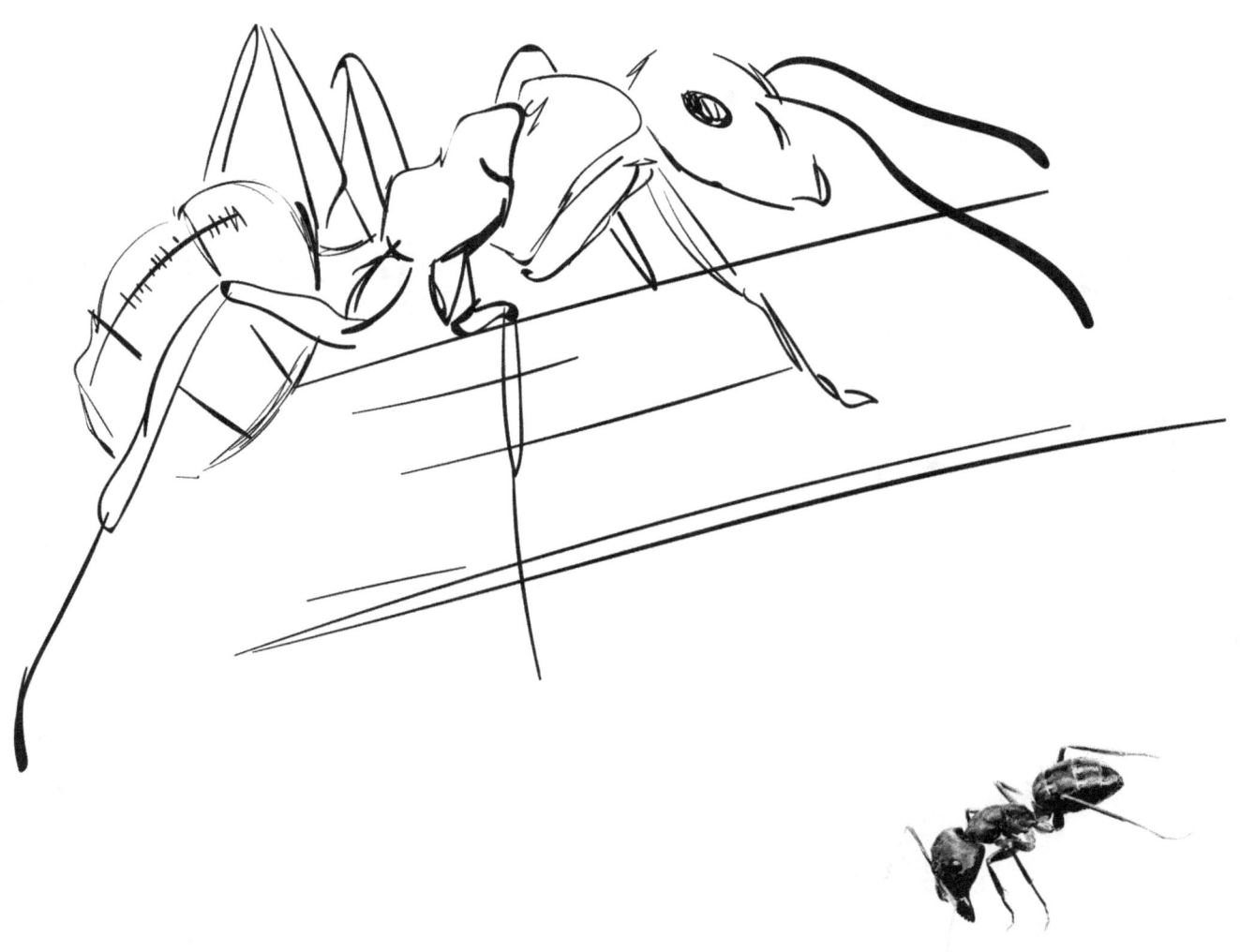

MEGAPONERA ANALIS

Megaponera analis ants can lose legs or even their heads when they fight termites

Megaponera analis

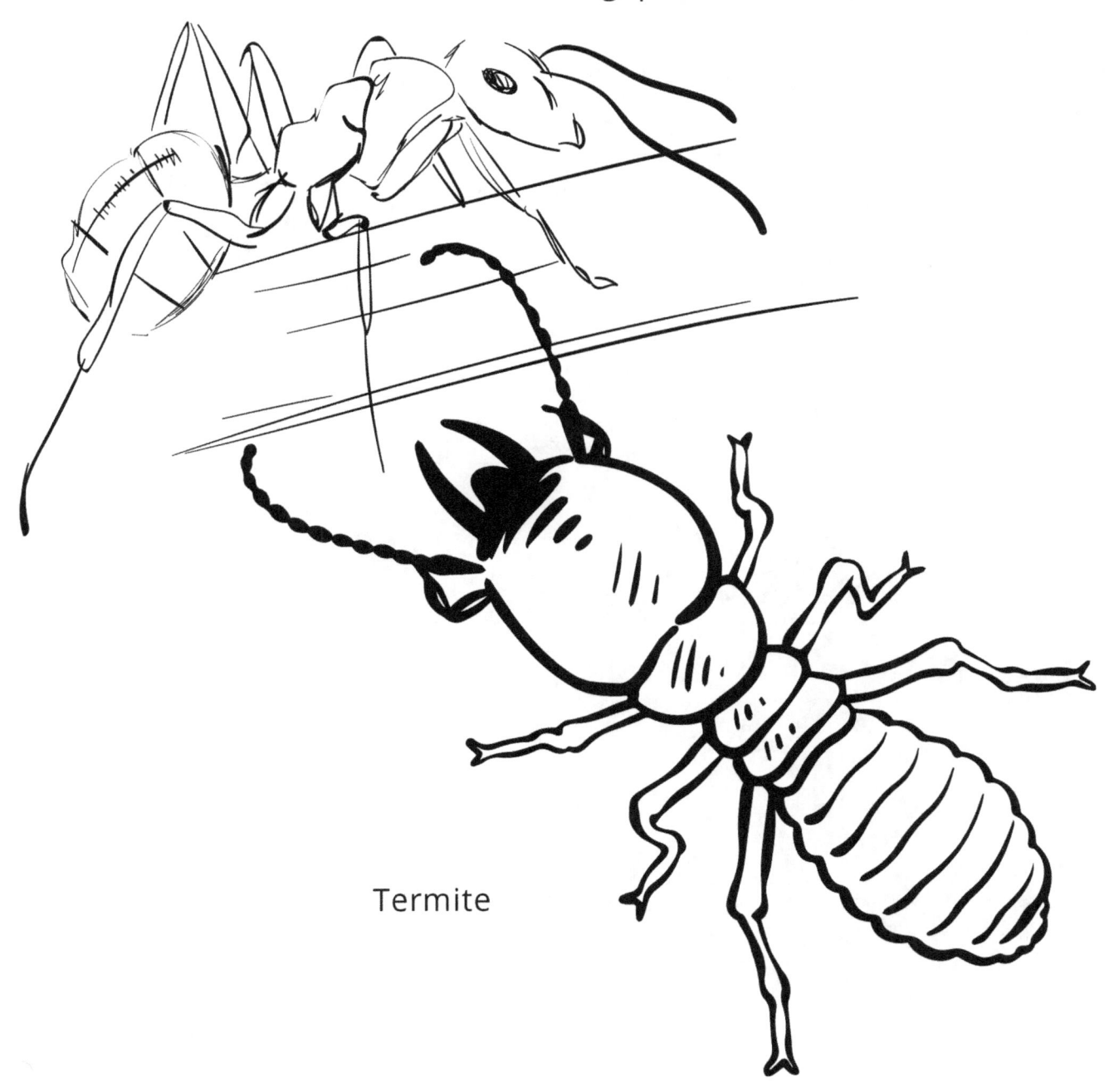

Termite

MEGAPONERA ANALIS

When Megaponera analis ants get hurt, their friends don't leave them behind. They carry them back home to heal so they can help again in future battles!

ANT EATER

Anteaters really do eat ants - it's their favorite food! In fact, one giant anteater can eat up to 30,000 ants in a single day using its long, sticky tongue

ANT POPULATION

There are 20 quadrillion ants on Earth, that's 2.5 million ants for every human.

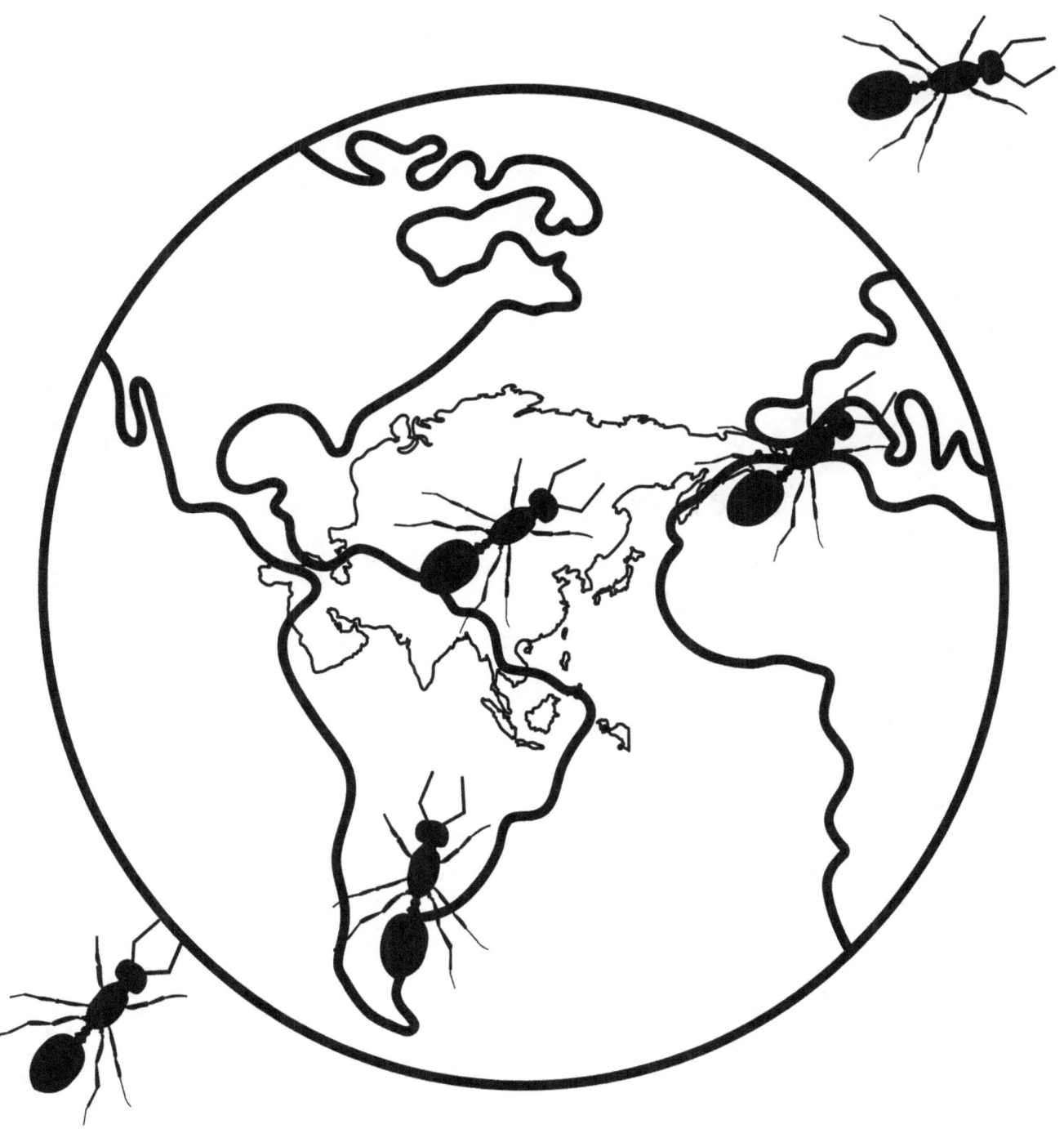

AMAZING ANTS

There are over 12,000 kinds of ants that scientists know about, and some think there might be more than 20,000 kinds!

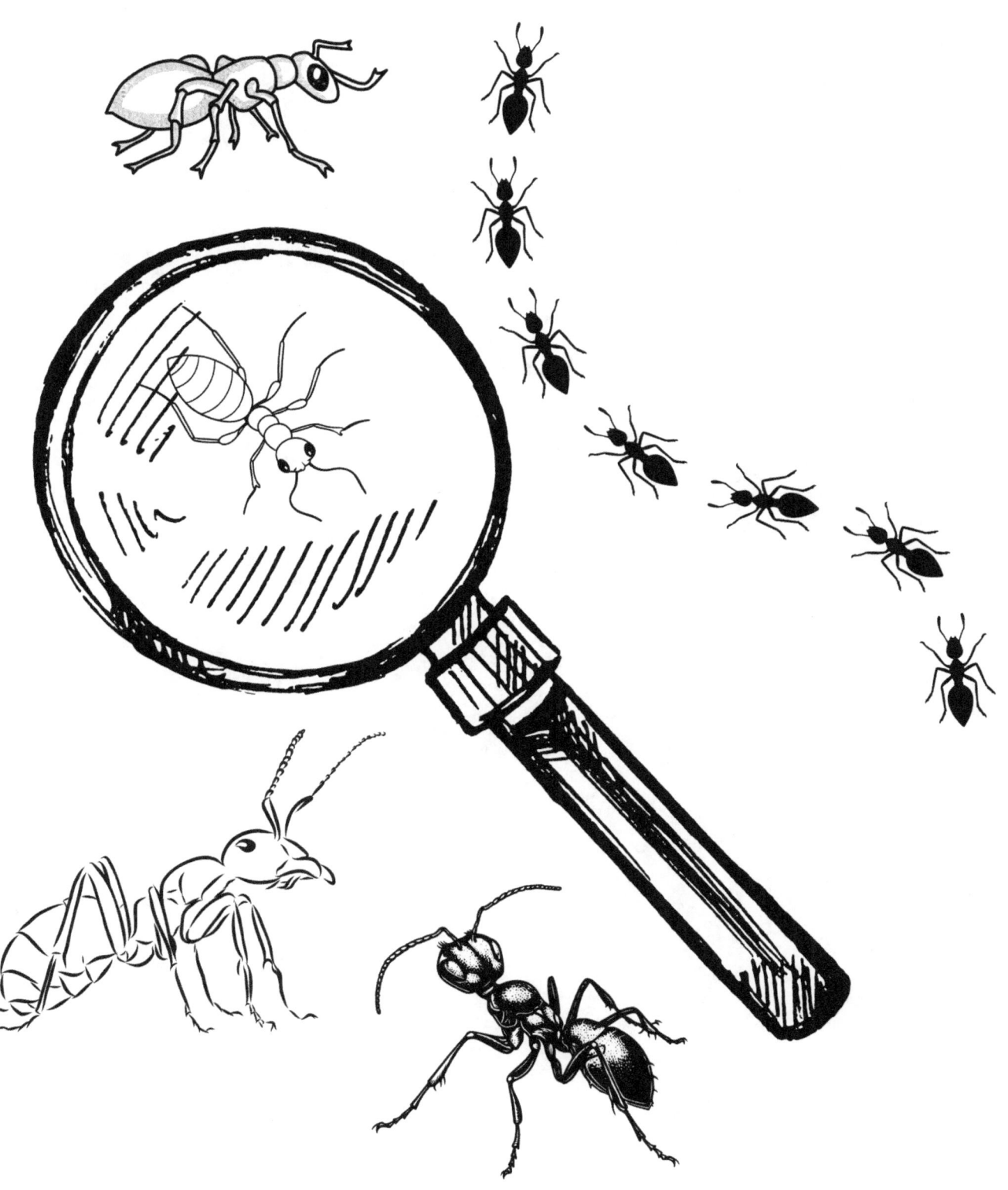

INVASIVE ANT SPECIES

Some ants are moving to new places where they don't belong, causing problems.

INVASIVE ANT SPECIES

Scientists found over 500 kinds of ants in places they weren't born, often traveling with people or in shipped packages!

INVASIVE ANT SPECIES

Alien ants - ants that don't belong in a place - can change how an ecosystem works. An ecosystem is all the plants, animals, and insects living together in one area.

INVASIVE ANT SPECIES

When alien ants take over, they eat a lot of food and can hurt or chase away other animals and plants, making it hard for the whole neighborhood of nature to stay healthy!

ANT ECONOMICS

Ant economics is the study of how ant colonies manage resources, divide labor, and make decisions - just like a tiny, well-run society or business!

ANT ECONOMICS

Ants take risks to find food. Some explore dangerous places, but if it pays off, the whole colony benefits. It's like investing in a new idea - there's a risk, but also a big reward if it works.

ANT ECONOMICS

Ant colonies can change plans fast if danger comes or food runs out. That flexibility helps them survive -like how businesses adapt to new markets or challenges.

DID YOU KNOW?

Ants make a sweet juice called "trophallaxis" that they share with their friends. It's like sharing yummy ant juice to help everyone stay healthy and strong!

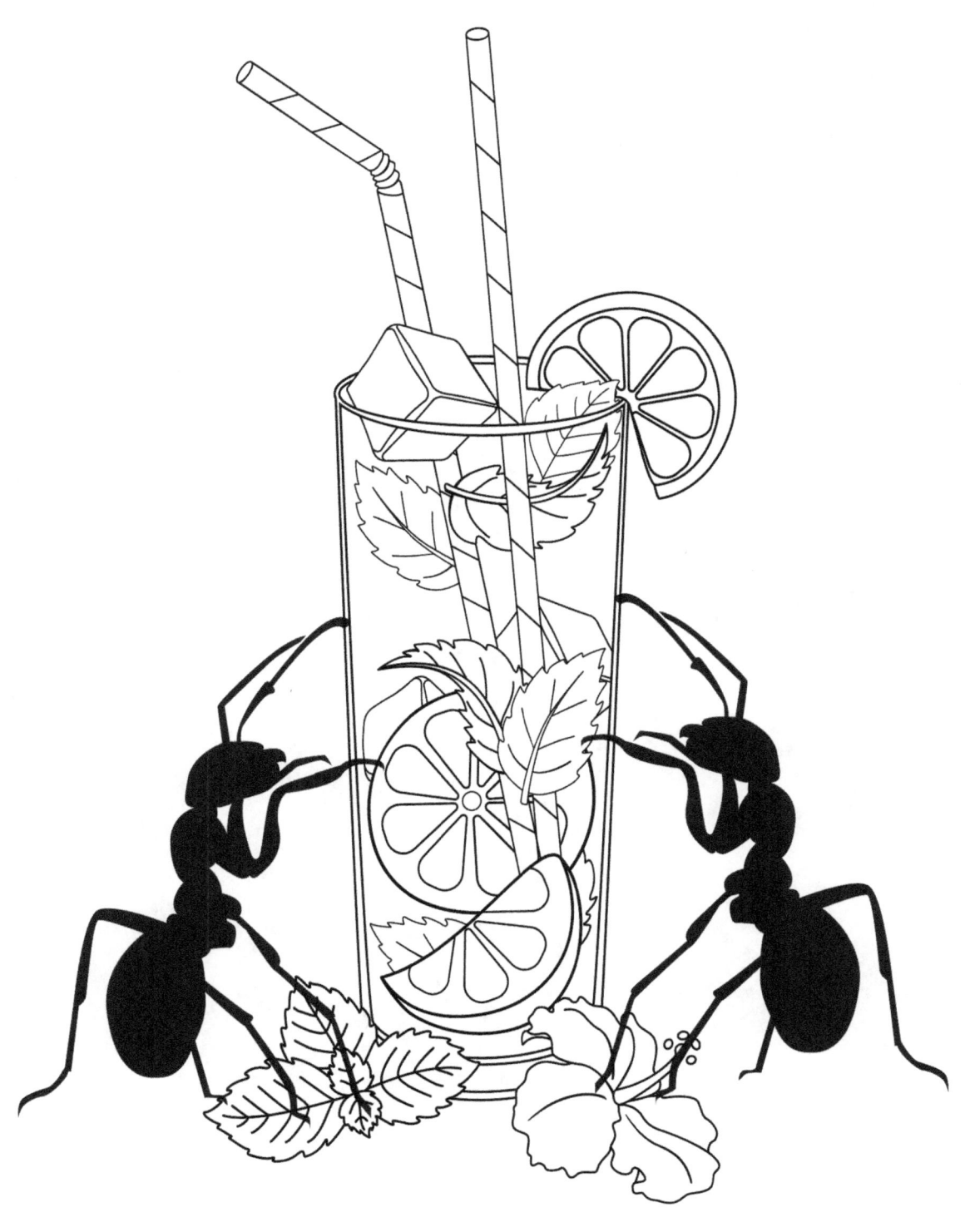

TINY FARMERS

Some ants grow their own food by farming fungus inside their nests.

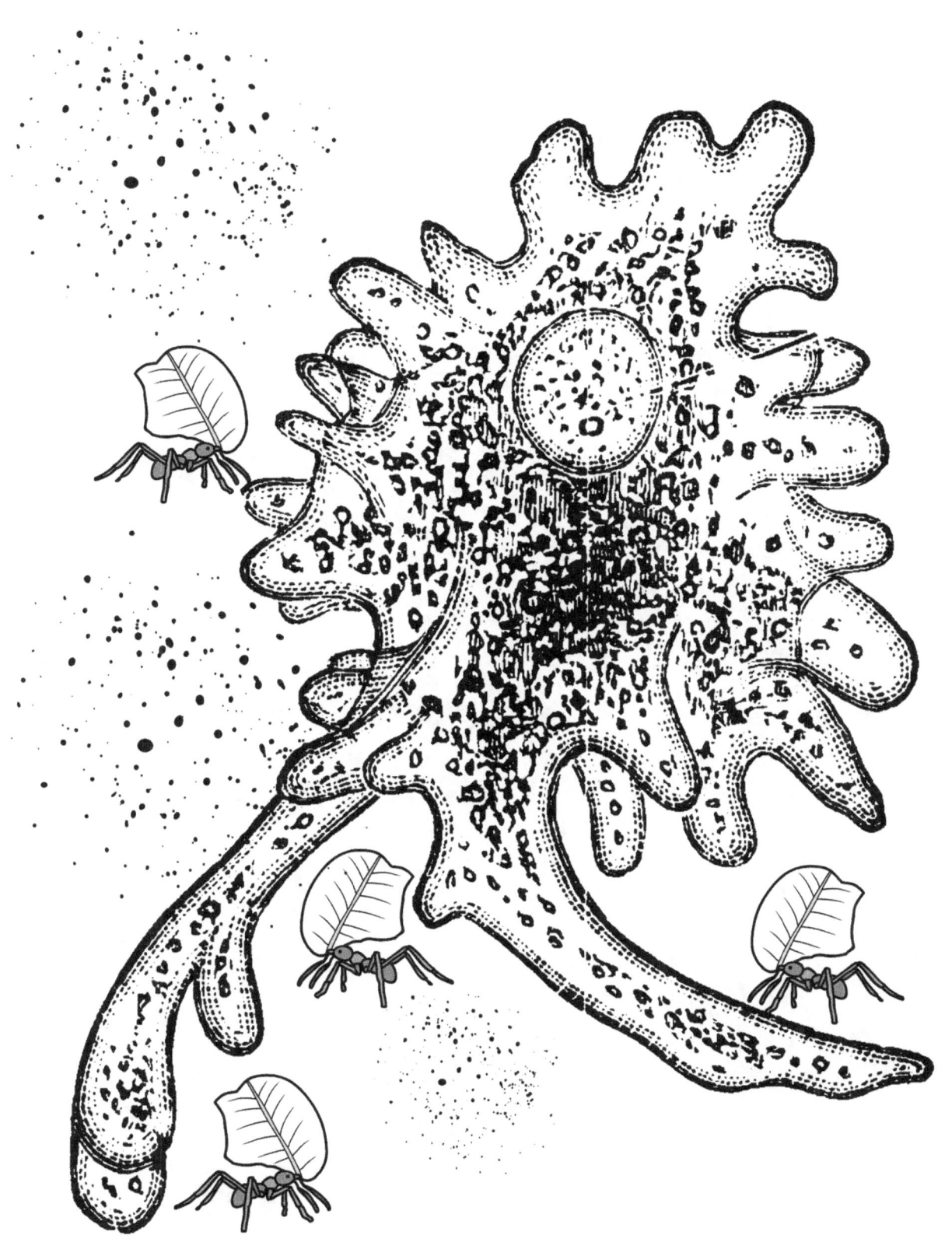

ANT BRIDGES

Army ants can link their bodies together to make bridges so other ants can cross!

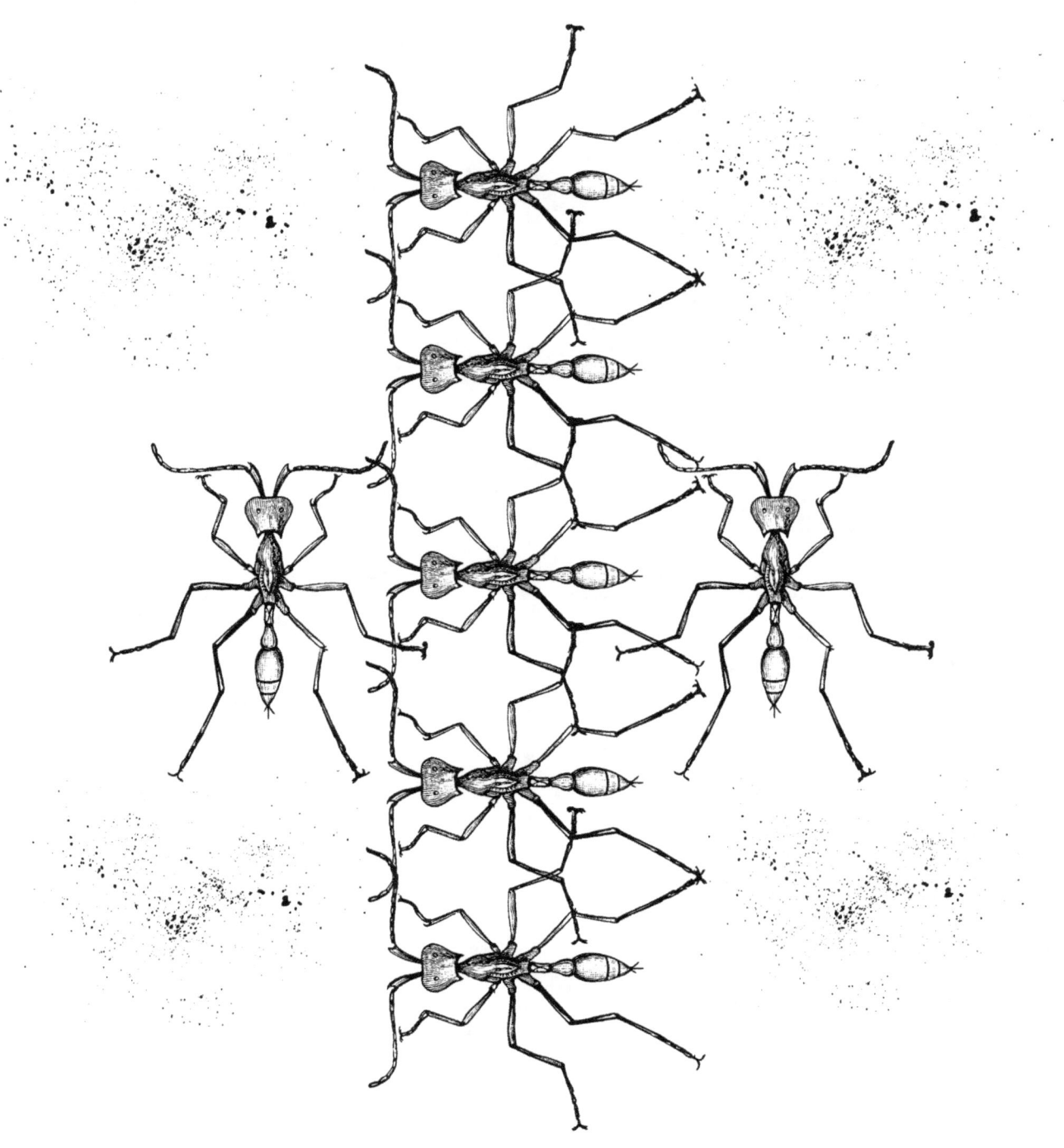

LONG LIFE

Queen ants can live for many years, sometimes over 30 years!

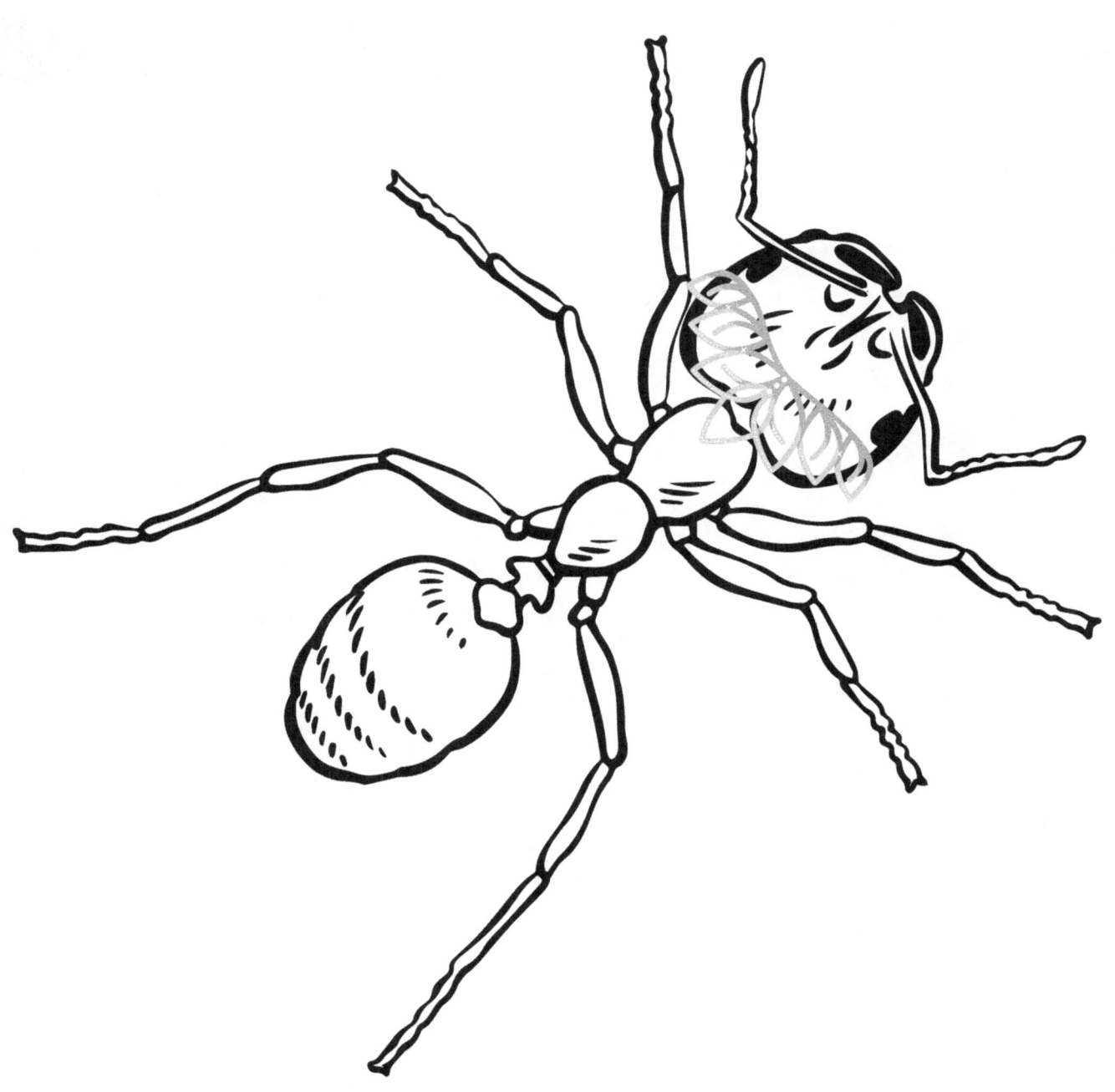

TINY BUILDERS

Ants are amazing architects! They build tunnels and chambers underground that can have many rooms, kind of like a little ant city

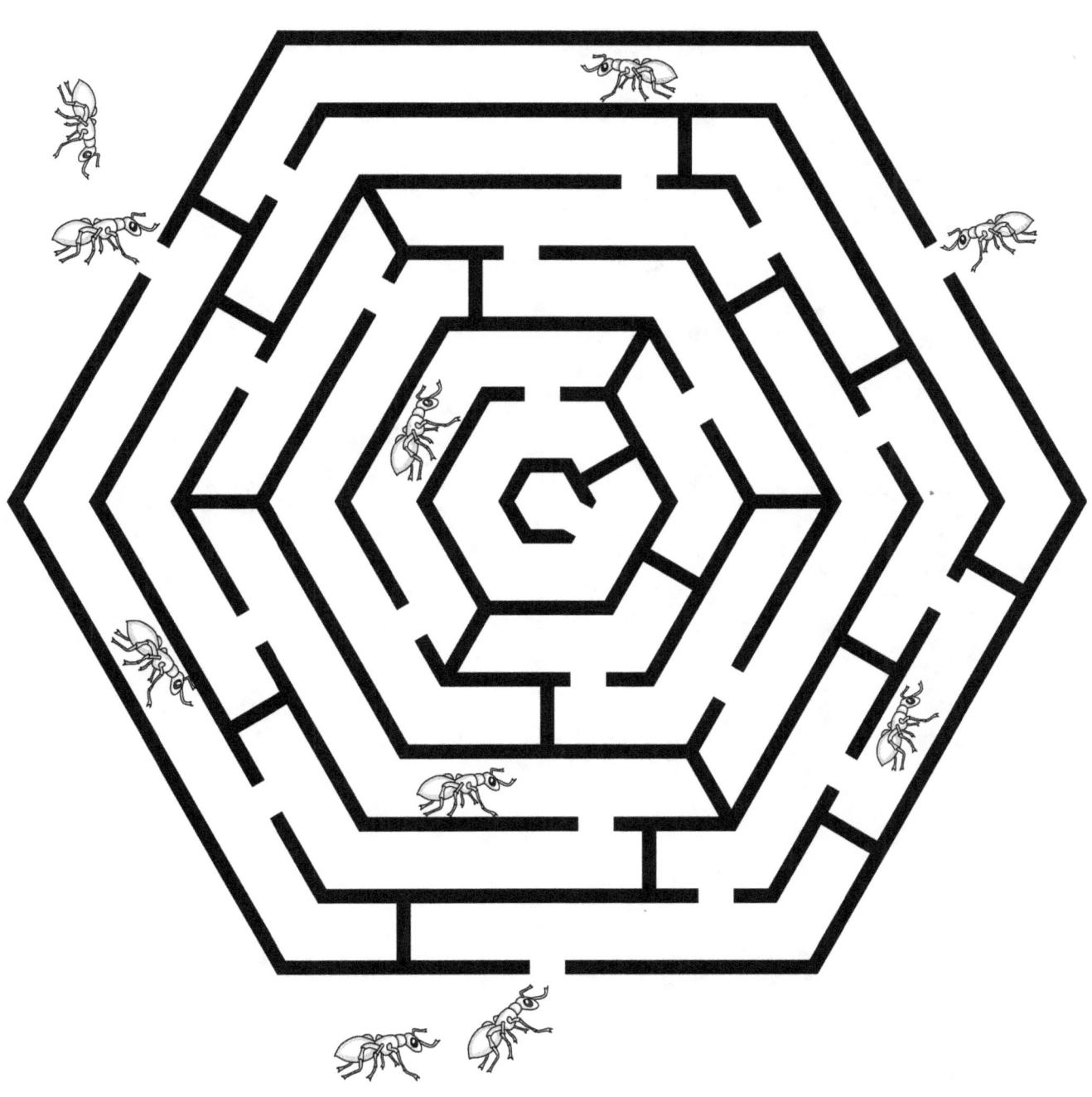

TEAMWORK

Ants never work alone-they always help each
other to find food, protect their home, and take
care of baby ants.

ANT-LETES

Some ants can run super fast-up to 3 miles per hour! That's speedy for such a tiny insect.

GLOBAL ANT

Ants live almost everywhere on Earth, except very cold places like Antarctica.

ANT SCOUTS

Some ants are scouts who go ahead to find food or new homes and then lead the others. They're like tiny explorers!

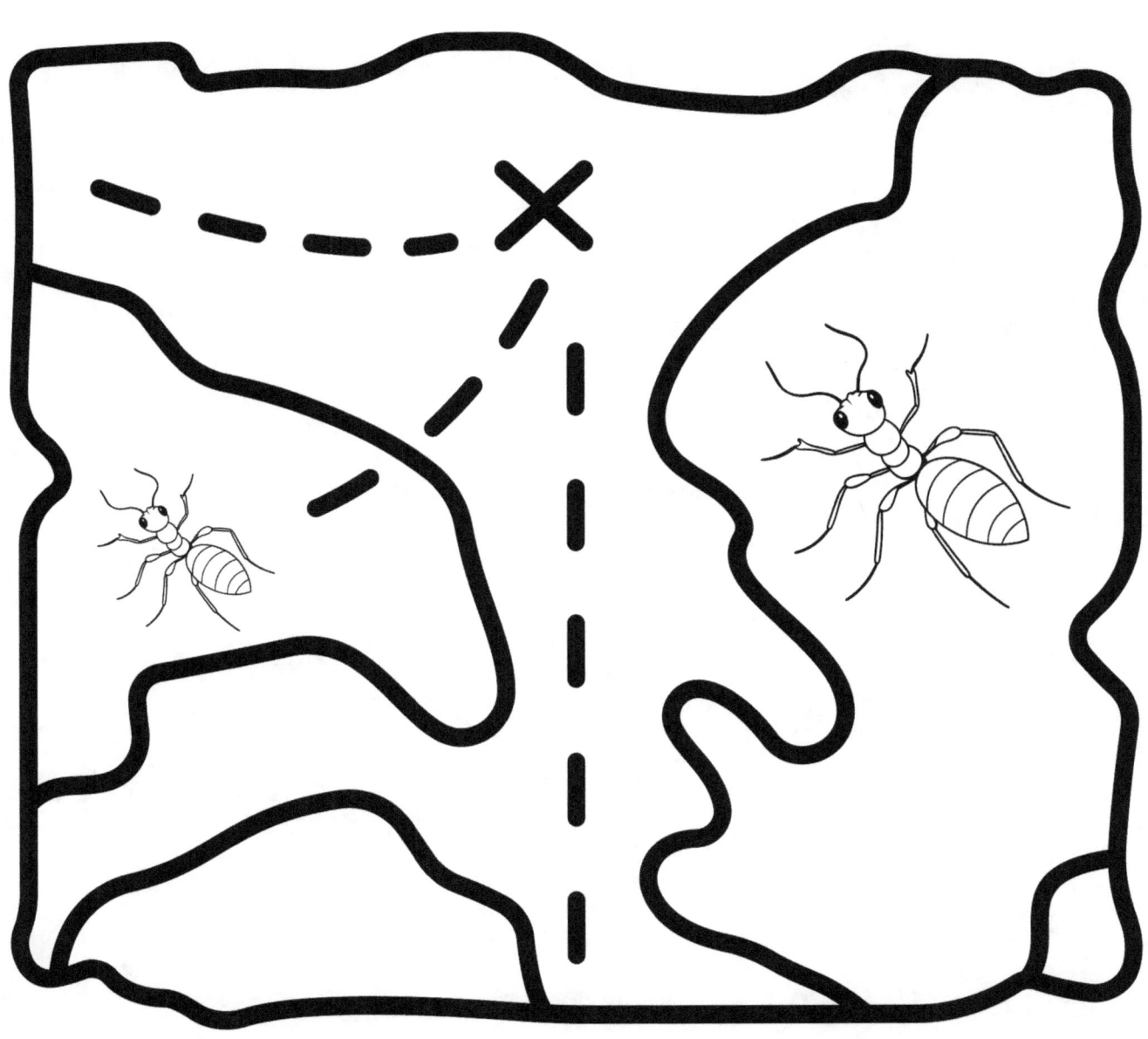

DID YOU KNOW?

Ants have two stomachs. One for eating and one just for sharing food with other ants.

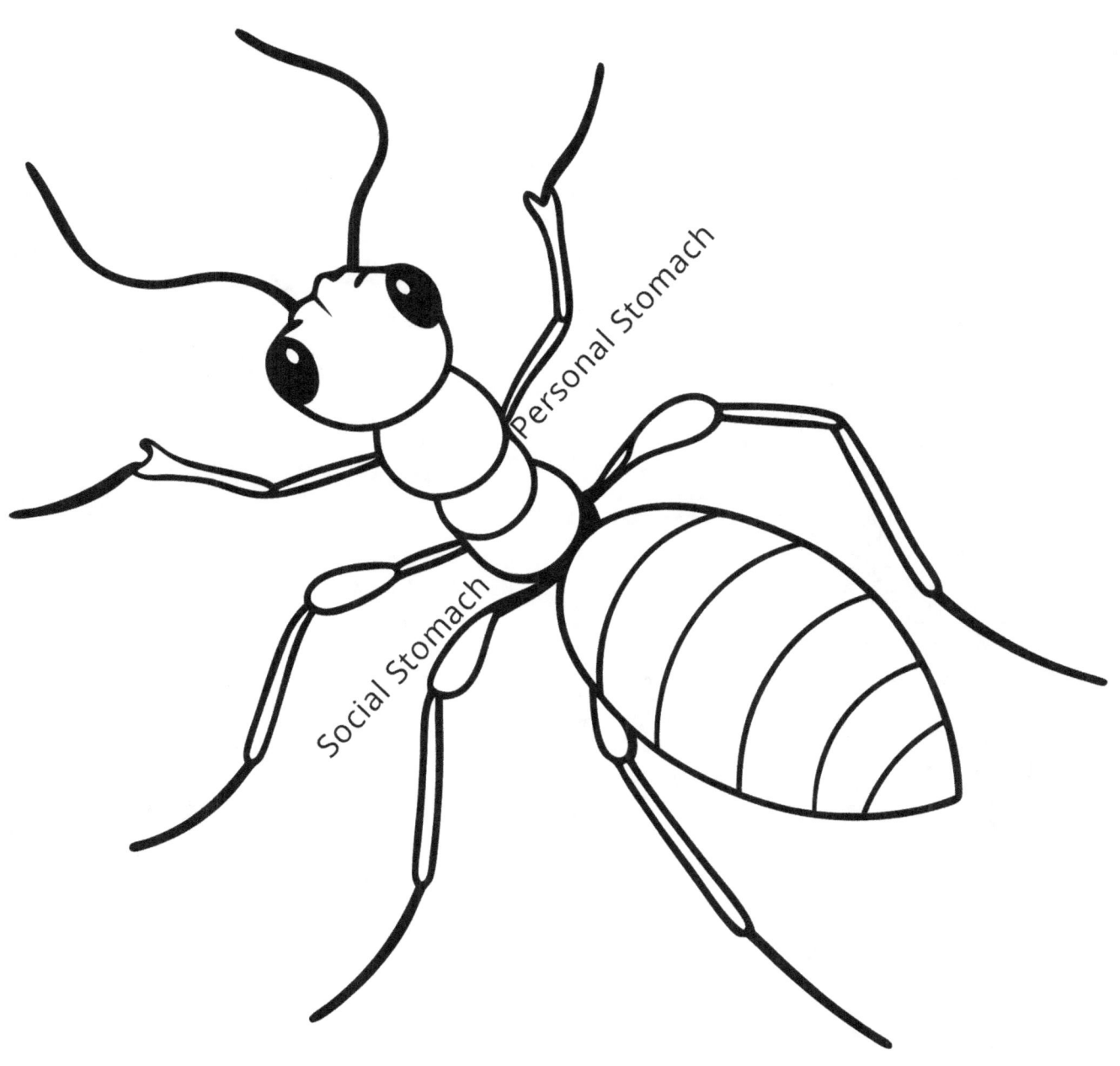

DID YOU KNOW?

Some ants can swim - they live near water and can float or swim to safety.

ANT FARM

Ants can "farm" aphids! They protect tiny bugs called aphids and collect the sweet honeydew they make.

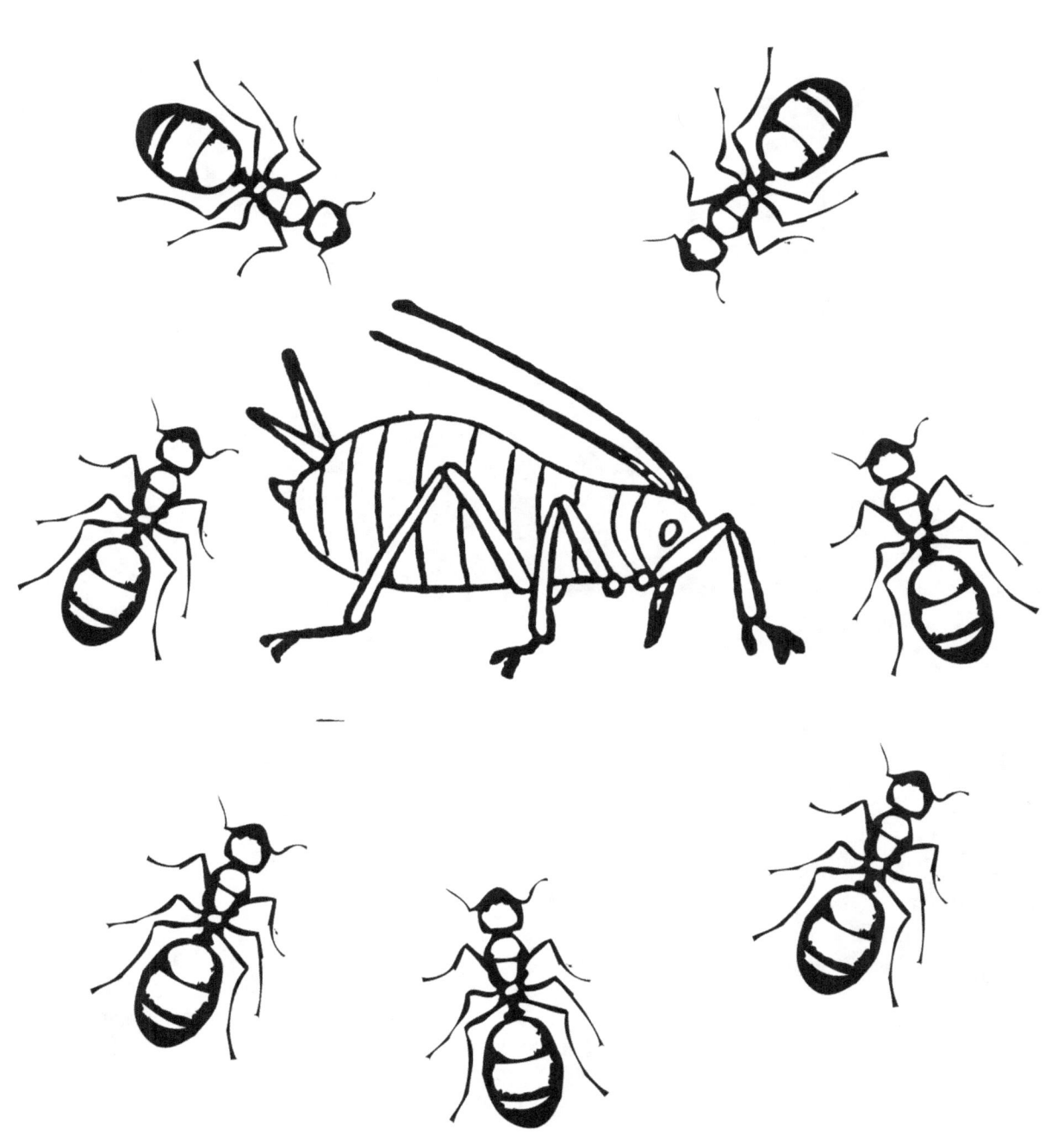

SOLDIER ANTS

These ants have big jaws and are tough fighters who guard their home from intruders.

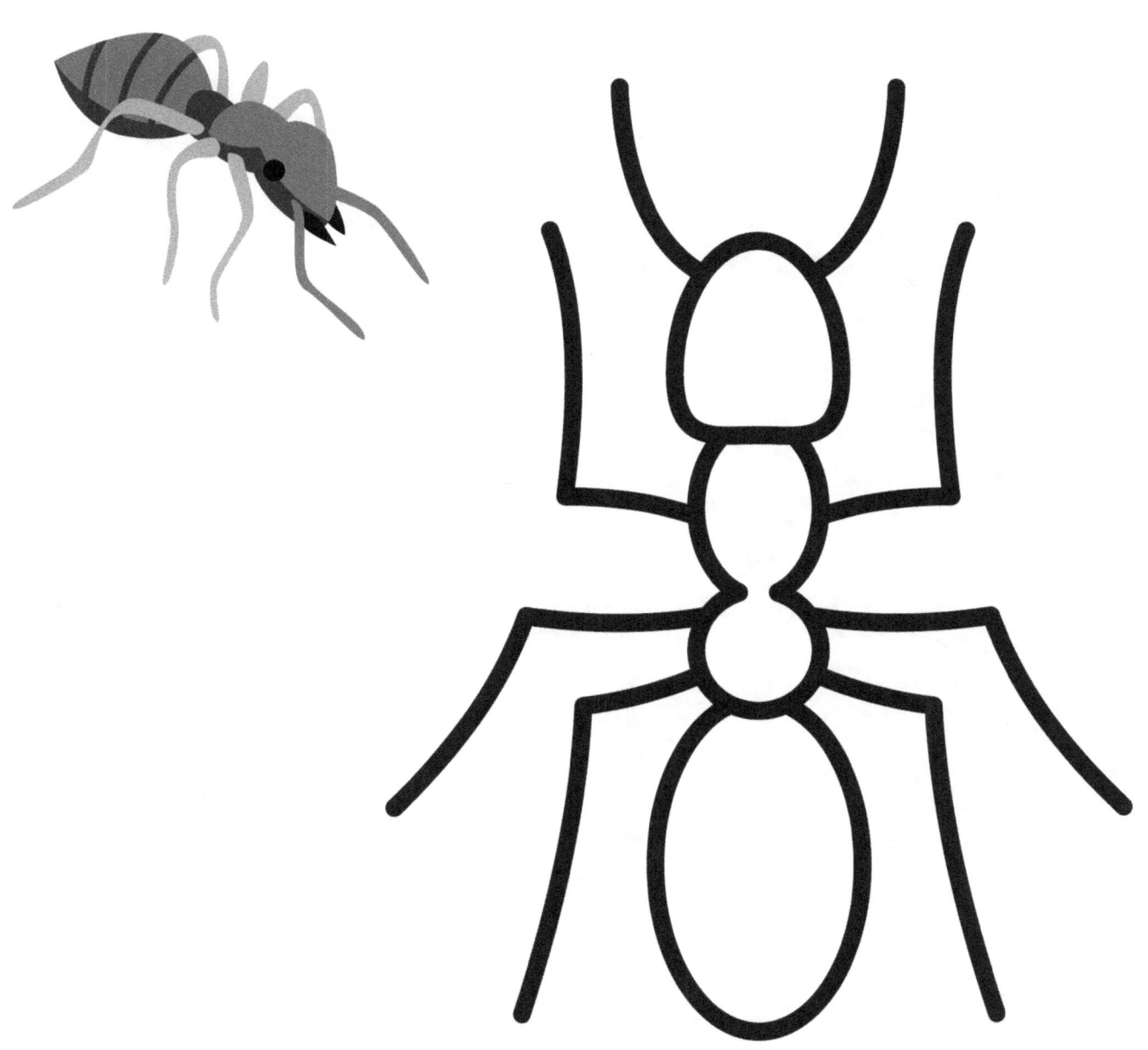

DID YOU KNOW?

Some ants live in trees - making nests in leaves or hollow branches. They never have to dig underground.

ANT TUNNELS

Ants dig big tunnels underground with rooms for babies, food, and the queen. Some tunnels have tiny air vents that keep the nest cool and fresh!

SCAREDY ANTS

When ants get scared, some can release a
smell that warns the whole colony!

PET ANTS

Some people keep ants as pets!

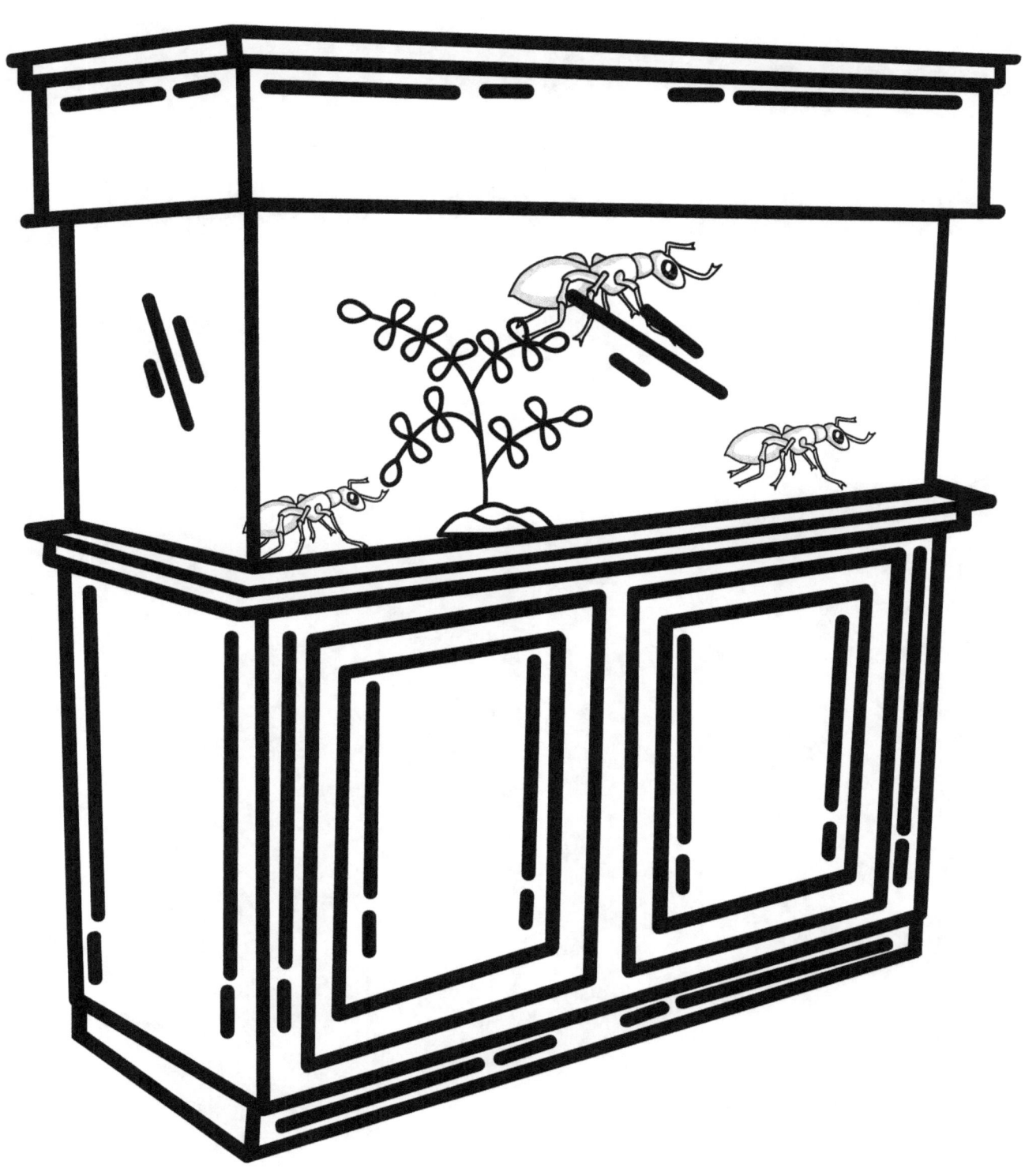

PET ANTS

Some known pet ants include the seed-collecting ants, sidewalk ants, and even honeypot ants that hold nectar in their big round bellies to feed their friends later!

HELL ANT

Scientists found the oldest ant ever-113 million years old! It's called a "hell ant" and had sharp, scythe-like jaws!

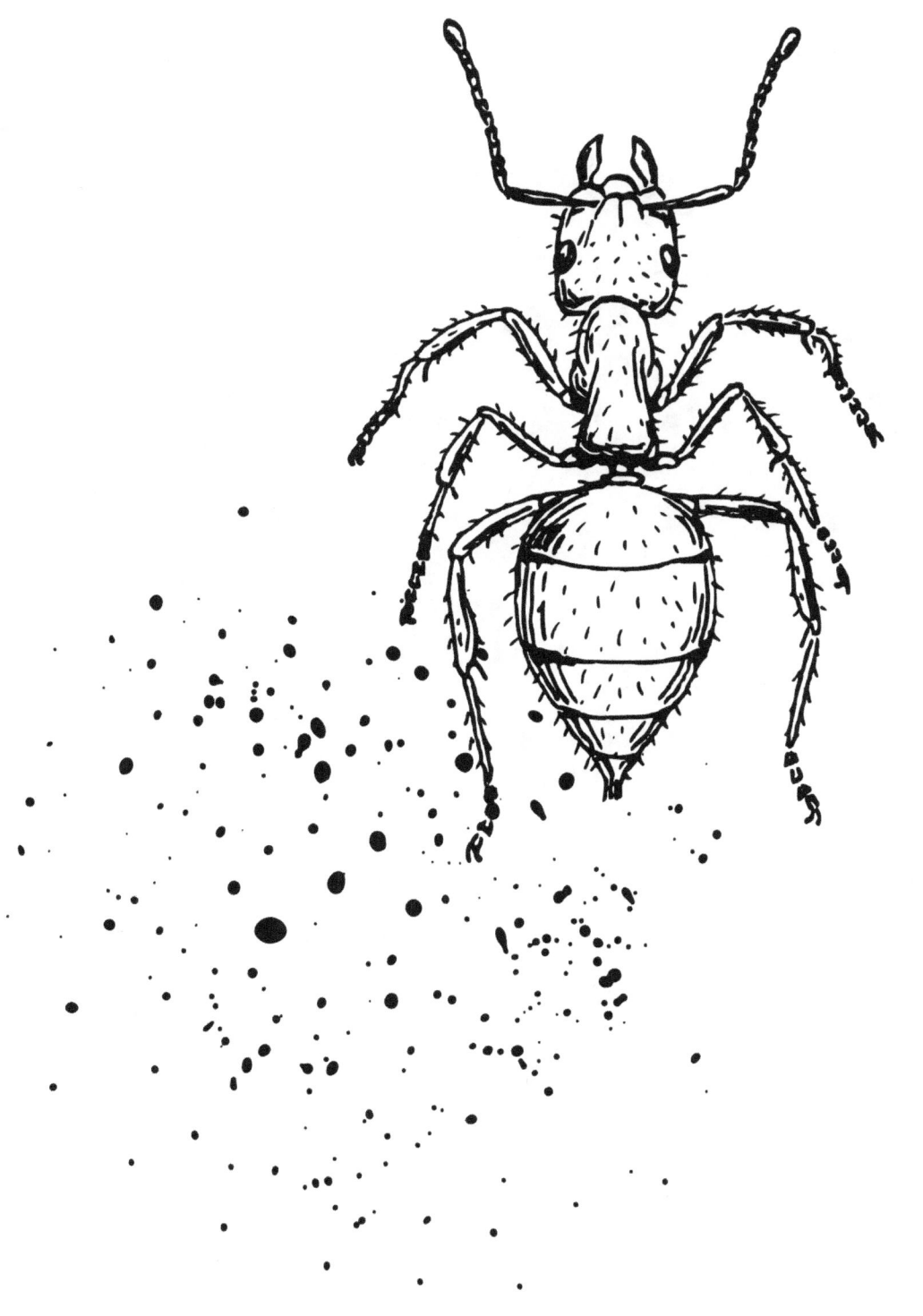

BUTTERFLIES AND ANTS

Some blue butterfly caterpillars make sweet juice that ants love. In return, the ants protect them from danger - it's a special friendship called a mutualistic relationship!

AUSTRALIAN BULL ANT

Australian bull ants go out at night to find tree sap and other snacks. New research shows they can use moonlight to find their way home.

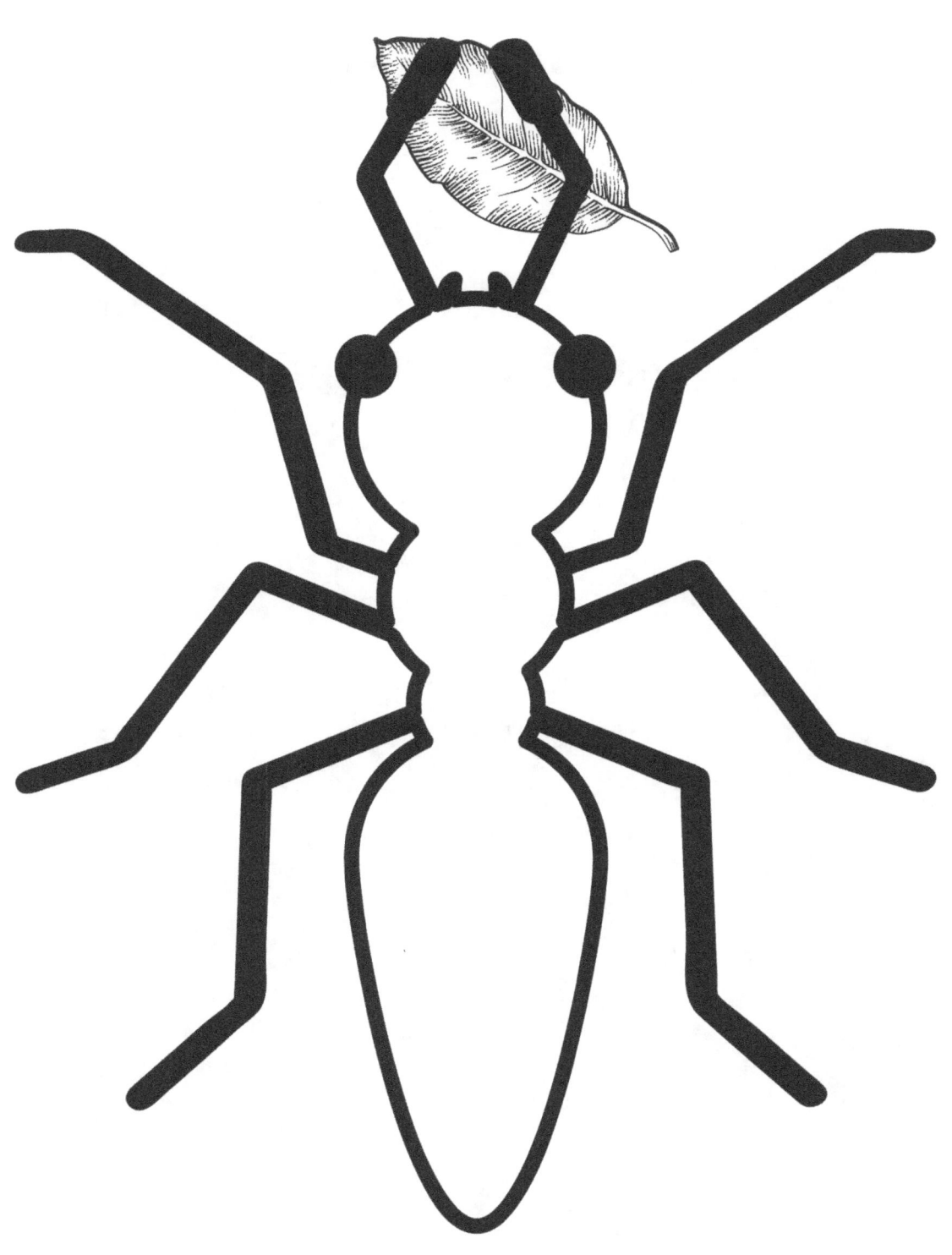

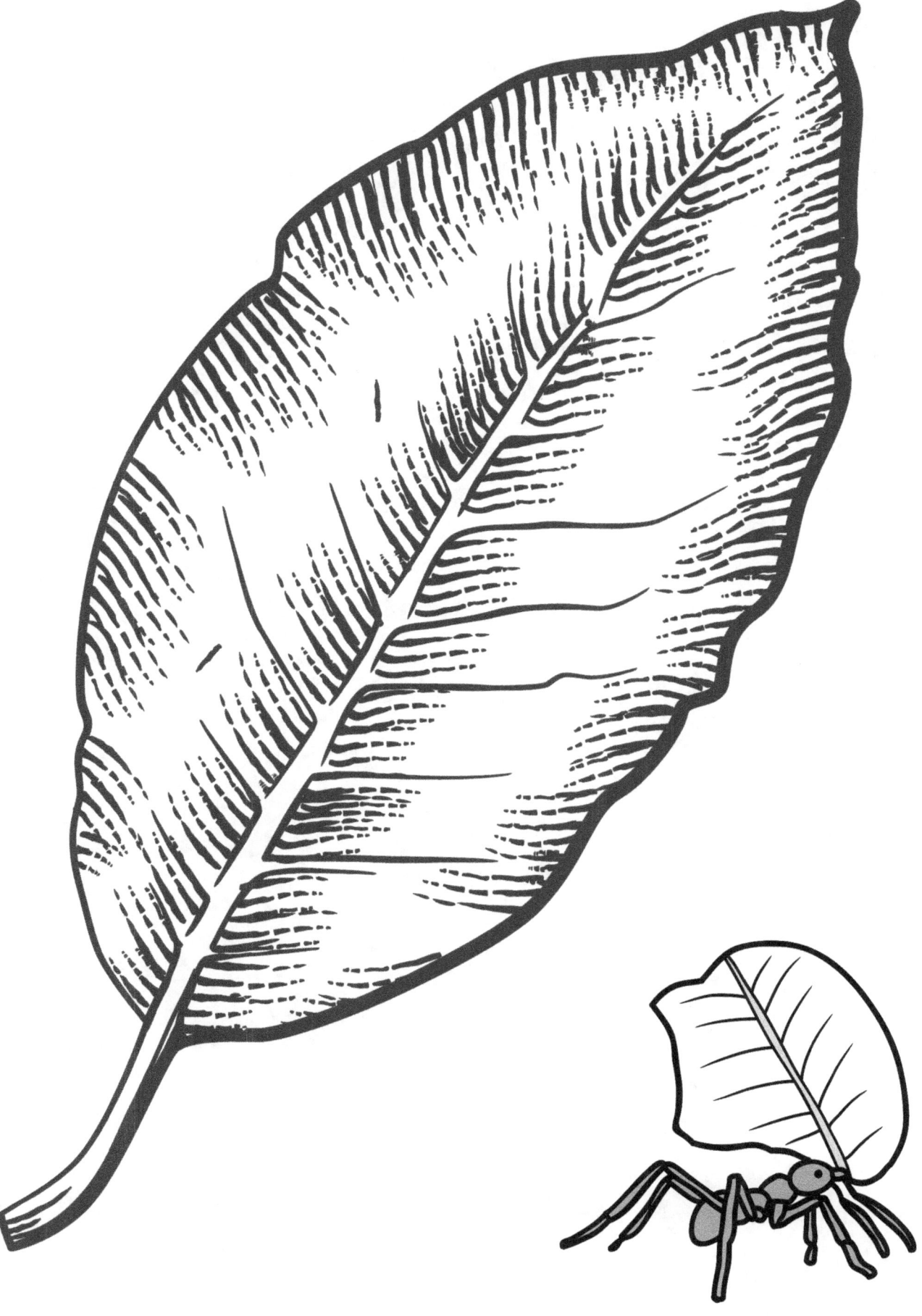

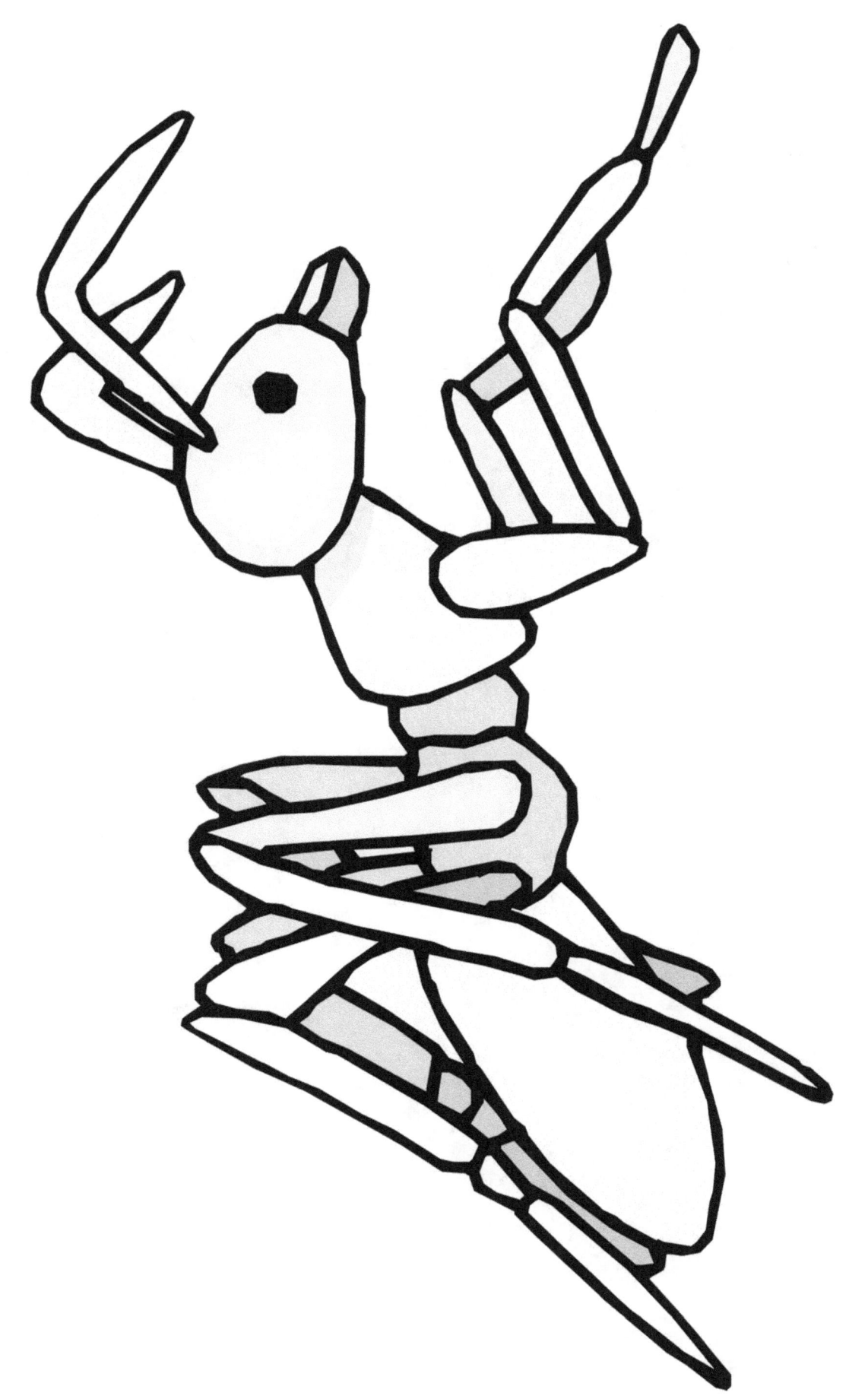

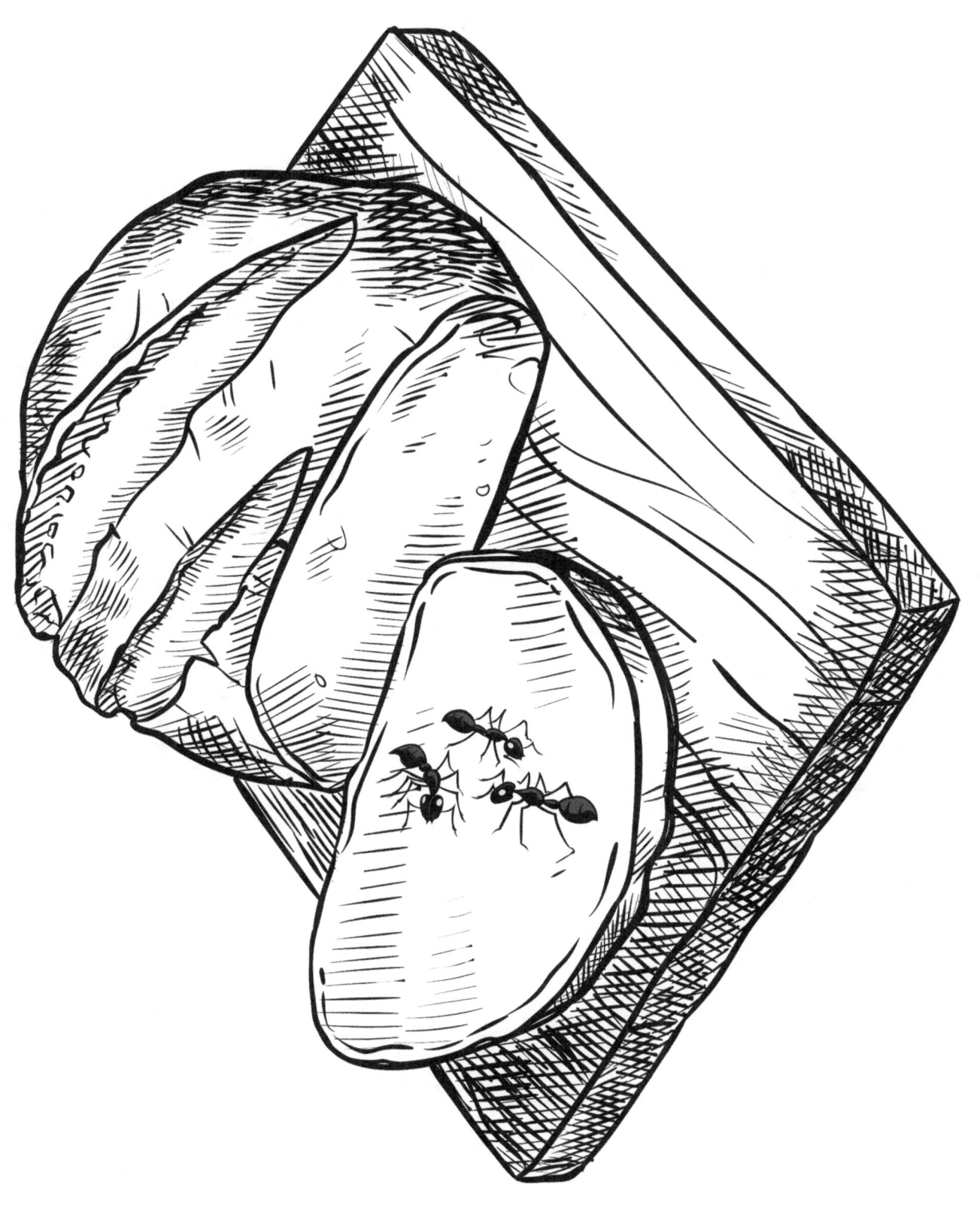

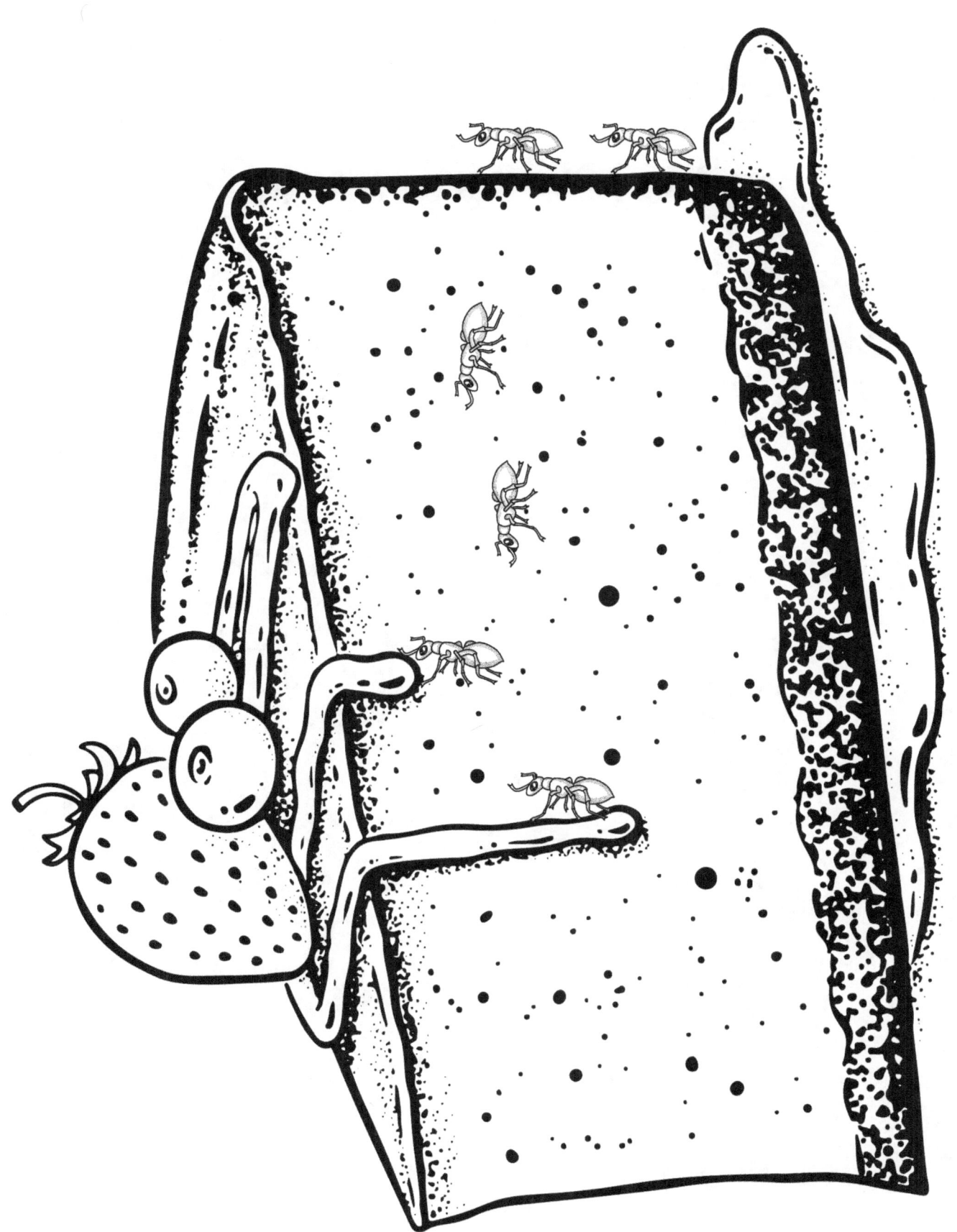

NOTES